Pharmacological Basis of

Yoo Kuen Chan • Kwee Peng Ng
Debra Si Mui Sim

Editors

Pharmacological Basis
of Acute Care

 Springer

Editors
Yoo Kuen Chan
Department of Anesthesiology
Faculty of Medicine
University of Malaya
Kuala Lumpur, Malaysia

Kwee Peng Ng
Subang Jaya Medical Centre
Subang Jaya, Malaysia

Debra Si Mui Sim
Department of Pharmacology
Faculty of Medicine
University of Malaysia
Kuala Lumpur, Malaysia

ISBN 978-3-319-35563-4   ISBN 978-3-319-10386-0 (eBook)
DOI 10.1007/978-3-319-10386-0
Springer Cham Heidelberg New York Dordrecht London

Illustrator: Mr Loke Lee Wong

Printed on acid-free paper

Springer is part of Springer Science+Business Media (www.springer.com)

Loke Lee, Hsui Han and Hsui Yang, thank you for not only giving me the time and space to find myself in every endeavor I am involved in, but the boundless energy and creativity to enjoy the process. . . Yoo Kuen

Possible because of a family that enabled me to achieve whatever I want and a friend who never gave up. . .Kwee Peng

To my late Mum, Madam Goh Ghee Kiam – an ordinary woman with an extraordinary heart, who gave me the opportunity of a lifetime to further my study overseas so that I can be where I am now . . . Debra

Foreword

Physicians and others, practicing 'healing arts' have sought ways of curing symptoms and diseases by the administration of some form of 'potion' since the dawn of man. The application of herbs, animal parts and other naturally occurring substances has been a process of trial and error which improved as early prescriptions were written down and compared.

The drugs available for use are now very varied and knowledge of their exact pharmacological action and interaction continues to develop rapidly. All physicians struggle to keep up to date and the appearance of a new and helpful textbook to aid them is most welcome.

Anaesthesiologists pride themselves in being expert clinical pharmacologists in the hospital setting. This expertise has been acquired through years of acute clinical practice founded on a sound grounding in pharmacology which has always been at the forefront of our discipline. In anaesthesia, as in many fields of acute care, the needs of the patient are constantly changing. If we are able to meet this challenging task of acceding to their varying pharmacology/physiological demands, we have reached a very high level of practice with pharmacology.

This book represents the collective effort of a set of anaesthesiologists together with their pharmacology colleagues to bring these valuable concepts to those who truly need them in a simple to understand way.

World Federation of Societies of Anaesthesiologists   David J Wilkinson
London, UK

Preface

Pharmacology as taught to medical students and junior frontline providers has always been considered a subject difficult to grasp. Pharmacology textbooks and formularies are usually drug-centred. While clinical textbooks and treatment guidelines are disease-centred and provide treatment recommendations, they seldom discuss why these treatments are chosen. In addition, the nature and extent of the subject makes it difficult to be captured into a single book to meet the needs of providers. Gaps in knowledge occur in frontline providers as a result and this translates into gaps in care.

The contributors of this book realize the importance of closing this gap. We have focused on the pharmacological principles as they apply in the different clinical situations and highlight how they are best utilized in managing the ill patient. Whilst information about drugs is relevant in every pharmacology book, we have decided to pare it down to the minimum here, so that readers will not be distracted from the main purpose of this book. We hope to make them understand that while drugs administered are always with the aim of correcting or improving a physiological process in the body, the physiological changes and pathology in patients, particularly the acutely ill ones, can alter the effects of drugs and the body's responses to these drugs, sometimes dangerously. Ultimately, frontline providers should be able to use knowledge of pharmacology to provide maximum benefit and minimum harm to our acutely ill patients, who are most at risk of adverse drug events.

As pharmacology is second nature to all the authors, the majority of whom are practicing clinicians, we hope to share our wealth of experience to make this vast knowledge available to providers in a way that is as realistic and clinically relevant as possible. This will hopefully shorten the learning curves of those who strive to do their best in the most difficult of conditions, i.e. to master the proper use of medication and drug resources to allow the body to heal.

We have also taken the extra step of including information applicable to acute care situations where appropriate decisions must be made within short time frames. Equipping oneself with the correct principles in this aspect of pharmacology will allow the provider to manage these critically ill patients effectively and safely.

Kuala Lumpur, Malaysia   Yoo Kuen Chan
Subang Jaya, Malaysia   Kwee Peng Ng
Kuala Lumpur, Malaysia   Debra Si Mui Sim

Contents

Contributors

Lucy Chan, M.B.B.S., FANZCA Department of Anesthesiology, Faculty of Medicine, University of Malaya, Kuala Lumpur, Malaysia

Yoo Kuen Chan, M.B.B.S., FFARCS Department of Anesthesiology, Faculty of Medicine, University of Malaya, Kuala Lumpur, Malaysia

Sook Hui Chaw, M.D., M.Anaes Department of Anesthesiology, Faculty of Medicine, University of Malaya, Kuala Lumpur, Malaysia

Kit Yin Chow, M.B.B.S., M.Anaes Department of Anesthesiology, Faculty of Medicine, University of Malaya, Kuala Lumpur, Malaysia

Li Lian Foo, M.D., M.Anaes Department of Anesthesiology, Faculty of Medicine, University of Malaya, Kuala Lumpur, Malaysia

Mohd Shahnaz Hasan, M.B.B.S., M.Anaes Department of Anesthesiology, Faculty of Medicine, University of Malaya, Kuala Lumpur, Malaysia

Noorjahan Haneem Md Hashim, M.B.B.S., M.Anaes Department of Anesthesiology, Faculty of Medicine, University of Malaya, Kuala Lumpur, Malaysia

Ahmad Khaldun Ismail, M.B.B.Ch., B.A.O., B.Med.Sc., Dr.Em.Med Department of Emergency Medicine, Faculty of Medicine, UKM Medical Centre, Kuala Lumpur, Malaysia

Suresh Kumar, M.B.B.S., MRCP, PG Diploma Epid Hospital Sungai Buloh, Jalan Hospital, Sungai Buloh, Selangor, Malaysia

Hou Yee Lai, M.B.B.S., M.Anaes Department of Anesthesiology, Faculty of Medicine, University of Malaya, Kuala Lumpur, Malaysia

Pauline Siew Mei Lai, B.Pharm., Ph.D. Department of Primary Care Medicine, Faculty of Medicine, University of Malaya, Kuala Lumpur, Malaysia

Pui Kuan Lee, M.D., M.Anaes Department of Anesthesiology, Faculty of Medicine, University of Malaya, Kuala Lumpur, Malaysia

Pui San Loh, M.B.B.S., M.Med. Anaes, FANZCA Department of Anesthesiology, Faculty of Medicine, University of Malaya, Kuala Lumpur, Malaysia

Marzida Mansor, M.D., M.Anaes Department of Anesthesiology, Faculty of Medicine, University of Malaya, Kuala Lumpur, Malaysia

Kwee Peng Ng, M.B.B.S., M.Anaes, FANZCA Subang Jaya Medical Centre, Subang Jaya, Malaysia

Yong-Kek Pang, M.D., MRCP Division of Respiratory Medicine, Department of Medicine, Faculty of Medicine, University of Malaya, Kuala Lumpur, Malaysia

Ina Ismiarti Shariffuddin, MBChB, M.Anaes Department of Anesthesiology, Faculty of Medicine, University of Malaya, Kuala Lumpur, Malaysia

Debra Si Mui Sim, B.Sc., Ph.D. Department of Pharmacology, Faculty of Medicine, University of Malaya, Kuala Lumpur, Malaysia

Sukcharanjit Singh Bakshi Singh, M.D., M.Anaes Department of Anesthesiology, Faculty of Medicine, University of Malaya, Kuala Lumpur, Malaysia

Choo Hock Tan, M.B.B.S., Ph.D. Department of Pharmacology, Faculty of Medicine, University of Malaya, Kuala Lumpur, Malaysia

Jeyaganesh Veerakumaran, M.B.B.S., MMed Anaes Department of Anesthesiology, Faculty of Medicine, University of Malaya, Kuala Lumpur, Malaysia

Ramani Vijayan, M.B.B.S., FFARCS, FRCA, FANZCA Department of Anesthesiology, Faculty of Medicine, University of Malaya, Kuala Lumpur, Malaysia

Kang Kwong Wong, M.B.B.S., M.Anaes Department of Anesthesiology, Faculty of Medicine, University of Malaya, Kuala Lumpur, Malaysia

Carolyn Chue-Wai Yim, M.B.B.S., M.Anaes Department of Anesthesiology, Faculty of Medicine, University of Malaya, Kuala Lumpur, Malaysia

Nur Lisa Zaharan, MBBCh, B.A.O., B.Med.Sc., Ph.D. Department of Pharmacology, Faculty of Medicine, University of Malaya, Kuala Lumpur, Malaysia

Part I
General Principles of Pharmacology and Pharmaceutics

Chapter 1
Why Drugs Are Administered

Yoo Kuen Chan and Debra Si Mui Sim

Abstract We are alive due to the many physiological processes in the body that sustain life in an integrated and controlled manner. When these processes are not working adequately, care providers resort to the use of drugs to try to improve them. At other times, these physiological processes need to be reversed to reduce damage to the functions of cells, tissues, organs or systems in the body. Drugs may also be used for diagnostic reasons to determine the presence of a disease condition in a patient. Increasingly, drugs are used for non-clinical reasons to enhance a physiological process, usually in sports, to provide a competitive edge to the user. The variability in drug requirements may stem from pharmaceutical, pharmacokinetic and pharmacodynamic differences, and all these must be taken into consideration in meeting the needs of individual patients. A database of use of drugs in a country indirectly reflects the health status and disease patterns of that country.

Keywords Physiological processes • Pharmaceutical variability • Pharmacokinetic variability • Pharmacodynamic variability • Database of drug use

Introduction

Many physiological processes keep us alive. These processes ensure our energy and oxygen needs are met and our waste products are removed so that our internal environment is within normal physiological parameters. These processes work harmoniously in an integrated manner so that all the organs function well to ensure survival and wellbeing.

When we are healthy and well, these physiological processes are under good control and the reserves are adequate for various levels of function. In many stages

Y.K. Chan, M.B.B.S., FFARCS (Ireland) (✉)
Department of Anesthesiology, Faculty of Medicine, University of Malaya,
50603 Kuala Lumpur, Malaysia
e-mail: chanyk@um.edu.my

D.S.M. Sim, B.Sc., Ph.D.
Department of Pharmacology, Faculty of Medicine, University of Malaya,
50603 Kuala Lumpur, Malaysia

© Springer International Publishing Switzerland 2015
Y.K. Chan et al. (eds.), *Pharmacological Basis of Acute Care*,
DOI 10.1007/978-3-319-10386-0_1

3

of life, these physiological processes may not be working adequately and we resort to improving them by the use of agents we normally call drugs. So drugs are administered for a variety of reasons.

Why Drugs Are Administered

Drugs are mainly administered by care providers with the purpose of enhancing a physiological process or to reverse physiological processes that are damaging to the functions of the cells, tissues, organs or systems in the body. Inotropes such as β_1-receptor agonist, dobutamine increase the heart rate and contractility, thus improving cardiac output. Antithyroid drugs like carbimazole block thyroid hormone (thyroxine) synthesis and are used to arrest damages caused by hyperthyroidism, which manifests with multiple organ signs and symptoms.

Often, drugs may be administered to prevent an abnormal physiological process from causing harm in the future. Patients who have undergone organ or tissue transplants from donors are given cyclosporine, an immunosuppressant, to prevent rejection.

Drugs are also administered as a diagnostic test. In the Tensilon test, edrophonium is used to diagnose myasthenia gravis at the bedside. Edrophonium is a short-acting anticholinesterase agent which increases the concentration of acetycholine at the neuromuscular junction. A diagnosis of myasthenia gravis is made when the administration of edrophonium transiently improves muscle power.

Drugs may be used for non-clinical reasons too, to enhance levels of function of organs in the body. Steroids, amphetamines and even β_2-receptor agonists have been used, for example, to improve physical performance in sports. These may be considered a misuse of the drug, especially if it is used to enhance a physiological process to provide the user with a competitive edge in sports.

Drugs usually interact with receptors in the body in order to produce their intended or undesired effects, be they diagnostic, prophylactic, therapeutic, toxic, or even just purely recreational.

Meeting the Needs of Patients

Pharmacokinetic and Pharmacodynamic Variability

Drug therapy is done with the needs of patients in mind. Healthcare providers aim to maximize benefit and minimize risks. Individualizing doses to meet the needs of the patient is the hardest lesson to learn in pharmacology.

Every patient's dosage requirements are different and this variability (Table 1.1) stems from variations in how the patient's body handles the drug, which we term

Table 1.1 Pharmaceutical, pharmacokinetic and pharmacodynamic factors which determine how a patient ultimately responds to a drug

Pharmaceutical factor (prescription of drug regimen)	Pharmacokinetic factor (bringing drug to site of action)	Pharmacodynamic factor (drug effect/response)
Drug sources	Age, sex, weight	Level of functions of organs
Drug manufacturers	Compliance	Receptor status
Drug packaging/storage	Absorption	Genetic status
Dosage forms	Distribution	Tolerance
Dosage calculations	Metabolism	Interaction
Prescription	Elimination	
	Interaction	

"pharmacokinetics" as well as variations in how the drug interacts with the patient's tissues to produce its biological responses, which we term "pharmacodynamics".

In acutely ill patients, many of these parameters will not be known to the provider. An idea of the physiological status of these patients is obtained from their history, physical examination and investigations, allowing us to anticipate the likely response to the drugs prescribed. Most drugs in acute care are administered intravenously and hence can be titrated to effect. This is the most frequently used method to administer the drug until the patients' needs are met. The adequacy of response may then be determined through further monitoring of the responses expected of the drug as well as development of toxicity from the drug. Monitoring may be assisted by measuring the blood concentrations of the drugs.

Care providers administer drugs with the intention of enhancing the body's healing process. Most of these processes occur at the cellular or tissue level and hence drugs must reach these sites of action. Many variables determine the amount of drugs that ultimately arrive at these sites to produce the desired effect.

The principles of pharmacology assist the provider in making the most effective use of drugs. Some of these variables that have to be taken into consideration include the recommended dosing regimen, the route of administration, the compliance of the patient to the intake of the drug plus the absorption, distribution, metabolism and excretion of the drug. In addition, several physiological parameters in a patient influence the response to the drug, and these include the genetic make-up of the patient, drug receptor state, drug interactions and, with continued use of the drug, tolerance to the drug or rebound responses on withdrawal of the drug.

Most drugs in the acute care situation are administered intravenously, intramuscularly or subcutaneously. The absorption of most drugs is of course 100 % with the parenteral routes, but absorption of drugs administered orally may be incomplete. Even if the drug is totally absorbed through the gastrointestinal tract, it may undergo biotransformation in the liver before it gets distributed in the systemic circulation. This will result in a reduced systemic bioavailability of the drug (see Chap. 3).

Pharmaceutical Variability

Many factors may affect the drug even before it arrives in the pharmacy or whilst it is housed in the pharmacy. These include how the drug is manufactured, packaged, stored and transported to the dispensing site for distribution to patients.

Even after patients have received their medications from the pharmacy, how they keep the drugs at home may have an impact on the final therapeutic effects of these drugs.

Meeting the Health Needs of a Country

Interestingly, the utilization of drugs by a country indirectly reflects the health status of a country and can provide perspective on many aspects of the drug needs of the population. In addition it can provide information of choices available, the effectiveness of drugs available, morbidity associated with drug use and generally the pattern of drug needs in a country.

Key Concepts

- Drugs are administered to enhance or prevent certain physiological process from occurring at the cellular, tissue or organ level.
- Drugs administered must meet the needs of patients.
- Drug requirements in patients vary and this variability is caused by variability in the pharmacokinetics and pharmacodynamics of the drugs in the patients, as well as the pharmaceutical differences in the drugs concerned.
- Understanding the pharmacokinetic, pharmacodynamic and pharmaceutical principles of drugs, allow care providers to prescribe appropriately to meet the needs of patients.

Summary

Drugs are needed to enhance certain physiological processes in the body. Others may be needed to reverse abnormal processes or to prevent these from occurring. The administration of drugs is mainly for therapeutic purposes but may also be for prophylactic, diagnostic or even recreational needs of an individual. The appropriate dose of drugs for an individual varies due to pharmacokinetic, pharmacodynamic and pharmaceutical variability. Database on the drugs utilized by a country provides interesting perspective of the health status and disease pattern in a country.

Further Reading

1. Bapna JS, Tripathi CD, Tekur U. Drug utilisation patterns in the third world. Pharmacoeconomics. 1996;9(4):286–94.
2. Johannessen LC, Baftiu A, Tysse I, Valso B, Larsson PG, Rytter E, et al. Landmark Pharmacokinetic variability of four newer antiepileptic drugs, lamotrigine, levetiracetam, oxcarbazepine and topiramate: a comparison of the impact of age and comedication. Ther Drug Monit. 2012;34(4):440–5.
3. Levy G. Impact of pharmacodynamic variability on drug delivery. Adv Drug Deliv Rev. 1998;33(3):201–6.
4. Orchard JW, Fricker PA, White SL, Burke LM, Healey DJ. The use and misuse of performance-enhancing substances in sport. Med J Aust. 2006;184(3):132–6.
5. Sacristan JA, Soto J. Drug utilisation studies as tools in health economics. Pharmacoeconomics. 1994;5(4):299–312.
6. Seybold ME. The office Tensilon test for ocular myasthenia gravis. Arch Neurol. 1986;43(8):842–3.
7. Undevia SD, Gomez-Abuin G, Ratain MJ. Pharmacokinetic variability of anticancer agents. Nat Rev Cancer. 2005;5:447–58.

Chapter 2
Drug Administration

Debra Si Mui Sim

Abstract Many drugs do not make it from bench to bedside because of poor pharmacokinetic properties. Many medicines fail to save lives or achieve their optimal potentials because practitioners are ill equipped in their pharmacokinetic knowledge. Before a drug can produce its desired clinical effect, it must first be able to reach its target site, be it on the body surface or inside the body. The amount of a drug and the rate at which the drug reaches the target site depend in part on the route taken to administer the drug. The choice of the route of drug administration in turn depends on various factors which are related to therapeutic concerns and drug properties. The therapeutic concerns include questions relating to the desired onset rate and duration of drug action, where the drug target site is (readily accessible or not) and whether or not patient compliance is an issue. The drug properties to be taken into account include its physicochemical characteristics (e.g., lipid solubility, molecular size, ionization status) and plasma concentration-time profile. Drugs can be administered by a wide variety of routes, each with its advantages and disadvantages. Therefore making a right choice of route may be the start of a successful therapeutic intervention.

Keywords Pharmacokinetics • Routes of administration • Enteral routes • Parenteral injections • Route of choice • First-pass effect • Local or systemic effect

Introduction

Choosing a drug for treating a clinical condition is often a qualitative decision. It depends on the mechanism of drug action and the goal of the therapeutic intervention. Having made that qualitative decision, we must necessarily consider quantitative aspects to ensure that the drug can get to the target sites to produce the desired effect clinically. Many drugs do not make it from bench to bedside because of poor pharmacokinetic properties. Many medicines fail to save lives or

D.S.M. Sim, B.Sc., Ph.D. (✉)
Department of Pharmacology, Faculty of Medicine, University of Malaya,
50603 Kuala Lumpur, Malaysia
e-mail: debrasim@ummc.edu.my

© Springer International Publishing Switzerland 2015
Y.K. Chan et al. (eds.), *Pharmacological Basis of Acute Care*,
DOI 10.1007/978-3-319-10386-0_2

achieve their optimal potentials because practitioners are ill equipped in their pharmacokinetic knowledge. However, the amount of a drug and the rate at which the drug reaches the target site (be it on the body surface or inside the body) depend in part on the route taken to administer the drug. The choice of that route in turn depends on various factors which are related to therapeutic concerns and drug properties. Therefore making a correct choice of route may be the start of a successful therapeutic intervention.

What Is Pharmacokinetics

Before we consider the various routes of drug administration, let us first look at what pharmacokinetics is. Pharmacokinetics is the study of how drugs move within the body (Fig. 2.1). It describes the processes (rate and extent) by which an administered drug enters the body, gets distributed to the various tissues in the body and then eliminated from the body. There are two ways by which a drug is eliminated from the body: physically by excretion or chemically by metabolism (also known as biotransformation).

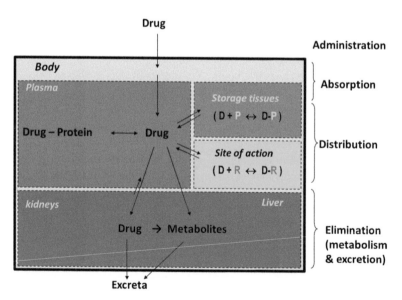

Fig. 2.1 Schematic diagram showing the various pharmacokinetic processes following administration of a drug. *D* drug, *P* protein, *R* receptor

How Pharmacokinetics Influence the Clinical Effect of a Drug

Pharmacokinetic processes, i.e., absorption, distribution, metabolism and excretion (ADME) determine how rapidly (hence, its speed of onset), in what concentration (hence, its intensity of effect), and for how long (hence, its duration of action) a drug will appear at its target tissues (Fig. 2.2).

Drug Administration

For a drug to be able to produce its intended clinical effects, it must first be able to reach its target site of action in the body at an effective concentration. If the drug is to act on some external surfaces, e.g., skin, ears or eyes, it may be applied directly on the affected surface. However, if the drug is meant to produce an effect inside the body, whether it be widespread (e.g., systemic antibiotics) or on some specific tissues (e.g., anti-thyroid agents), then the drug must be administered in such a way that it is able to get into the systemic circulation and be transported to the site (s) where the deranged body function is to be rectified.

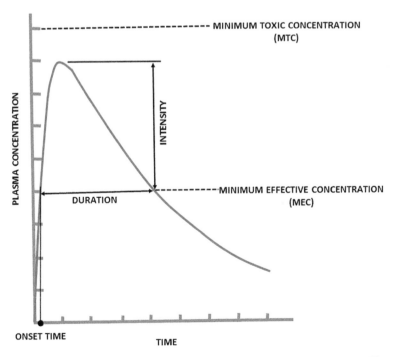

Fig. 2.2 Plasma concentration-time curve showing the onset time, duration of action and intensity of effects

How Drugs Are Administered

Drugs can be administered by a wide variety of routes. These may be generally divided into enteral routes (e.g., sublingual, oral, and rectal), parenteral injections (e.g., subcutaneous, intramuscular, and intravenous), inhalation, topical, transdermal and intranasal. Certain special routes may be used to provide better therapeutic outcomes: intra-arterial route in cancer chemotherapy and intrathecal route for central nervous system infections or for spinal anesthesia. These routes are much more hazardous and would obviously require greater skills and care in administering the drug.

Which Route of Administration

When making decision on the route of administration, two major factors come to mind: therapeutic concerns and drug properties. The therapeutic concerns include questions relating to the desired onset rate and duration of drug action, where the drug target site is (readily accessible or not) and whether or not patient compliance is an issue. The drug properties to be taken into account include its physicochemical characteristics (e.g., lipid solubility, molecular size, ionization status) and plasma concentration-time profile. The advantages and disadvantages of some of the common routes of administering drugs are given in Table 2.1.

Acute or Chronic Condition

In an acute or emergency situation, the route used must allow sufficiently fast absorption to ensure a prompt onset of action. Thus, the parenteral injection (intravenous, intramuscular or subcutaneous) is often the route of choice in acute care. Other routes that may be used in acute situations include intrathecal route for amphotericin B in cryptococcal meningitis, inhalation route for bronchodilators (by nebulizer) in acute bronchial asthma attack, or sublingual route for nitroglycerin in acute angina attack. With young children, the rectal route (as enema or suppository) may also be used, especially when the patient is unconscious or when vomiting occurs. However, for the treatment of chronic illnesses, the onset rate is not so much of a concern, whereas a more convenient route (e.g., oral, transdermal or topical) and simple dosing regimen (e.g., once-daily or once weekly dosing) would ensure better patient compliance. Formulation that can provide controlled release of the drug would be an added advantage especially in situation of poor patient compliance (e.g., severe depression). That is notwithstanding the need to take into account the cost factor. Transdermal route is usually more expensive because of the cost involved in manufacturing the delivery system.

Table 2.1 Advantages and disadvantages of some common routes of drug administration

Route	Advantages	Disadvantages
Intravenous (IV)	Absorption circumvented; prompt onset; Suitable for large volumes and for irritating substances	Most hazardous (embolism, infection, anaphylaxis); Not suitable for oily solutions or poorly hydrophilic substances
Intramuscular (IM)	Absorption may be tailored to needs: prompt, from aqueous solution; slow and sustained, from repository formulations; Suitable for moderate volumes, oily vehicles and some irritating substances	Precluded during anticoagulant therapy; May be painful
Subcutaneous (SC)	Absorption may be tailored to needs: prompt, from aqueous solution; slow and sustained, from repository formulations; Suitable for some poorly soluble suspensions and for instillation of slow-release implants; self-administration is acceptable	Not suitable for large volumes or irritating substances; Possible pain or necrosis from irritating substances
Oral (PO)	Most convenient; Relatively cheap and safe	Variable absorption (potentially slow, erratic and incomplete); First-pass effect may be significant
Rectal (PR)	Partially avoid first-pass effect; Avoid destruction by gastric acid & digestive enzymes	May irritate rectal mucosa; Not a well-accepted route
Sublingual (SL)	Prompt absorption; Bypasses first-pass effect (unless ingested)	Inconvenient for long-term use; Limited to certain types of drugs that can be given in small doses
Inhalation	Almost instantaneous absorption and very rapid onset; Avoid hepatic first-pass effect; May provide localized effect to lungs with minimal systemic side effect	Difficulty in regulating doses (inhaler); Requires special equipment for drug delivery
Transdermal	May provide a sustained effect; Avoid hepatic first-pass effect	Usually very slow onset; Enhanced absorption and risk of toxic effects with inflamed, abraded or burned skin; Drugs must be highly lipophilic

Lipophilic or Hydrophilic Drugs

The suitability of the route chosen also depends on the drug properties (both physical and chemical). Some drugs (e.g., theophylline, phenobarbital, propofol) are not sufficiently aqueous soluble to be given by intravenous injection per se and if needed, these drugs have to be formulated for better aqueous solubility (e.g., aminophylline, phenobarbital sodium) or given as an emulsion (e.g. propofol). On the other hand, most drugs are not sufficiently lipid soluble to be able to permeate the intact skin. Therefore, only highly lipid soluble drugs such as nitroglycerin, fentanyl, scopolamine, nicotine and estrogen can be formulated for sustained release by transdermal delivery system. Organophosphate insecticides are also very lipid soluble and poisoning can occur through contact with contaminated clothes.

High or Low First-Pass Effect

Some drugs are not stable in the gastric acid environment (e.g., benzylpenicillin) or may undergo extensive hepatic first-pass elimination (e.g., nitroglycerin) and are thus not suitable to be given orally. However some drugs, such as morphine and propranolol, may still be effective orally despite high first-pass effect because of active metabolites, which possess some of the pharmacological actions of the parent drugs.

Vaporized or Atomized Substances

Drugs which are gaseous (e.g., nitrous oxide) or readily vaporized (e.g., isoflurane) may be inhaled. Solid drugs may also be given by inhalation route in the form of aerosols or suspended powder (e.g., salbutamol and beclomethasone). The inhalation route may be used for producing both local (e.g., bronchodilatation in asthma) as well as systemic effects (anesthesia). It is also an important portal of entry for certain drugs of abuse and environmental toxicants. The main disadvantage with the inhalation route for therapeutic use is the difficulty in regulating the dose given and the cost of the delivery system. Intranasal route is usually used for local application (e.g., oxymetazoline for nasal decongestion) but it may occasionally be used for systemic effects (e.g., desmopressin for diabetes insipidus and calcitonin for osteoporosis).

Key Concepts

- For a drug to be clinically useful, it must also have an appropriate pharmacokinetic profile besides having the desired pharmacodynamic action.
- The choice of the route of drug administration depends mainly on two major factors: therapeutic concerns and drug properties.
- Parenteral injection routes are often the routes of choice in acute care while oral route is the most common route used in chronic illnesses.
- The physical and chemical properties of a drug can influence both the dosage form as well as the route(s) by which it may be administered.

Summary

The choice of the route of drug administration depends on various factors which are related to therapeutic concerns and drug properties. The therapeutic concerns include questions relating to the desired onset rate and duration of drug action, where the drug target site is (readily accessible or not) and whether or not patient compliance is an issue. The drug properties to be taken into account include its physicochemical characteristics (e.g., lipid solubility, molecular size, ionization status) and plasma concentration-time profile. Drugs can be administered by a wide variety of routes, each with its advantages and disadvantages. It is important to make the right choice in order to have a successful therapeutic outcome.

Further Reading

1. Lehman DF. Teaching from catastrophe: using therapeutic misadventures from hydromorphone to teach key principles in clinical pharmacology. J Clin Pharmacol. 2011;51(11):1595–602.
2. Lin JH, Lu AYH. Role of pharmacokinetics and metabolism in drug discovery and development. Pharmacol Rev. 1997;49(4):403–49.
3. Prentis RA, Lis Y, Walker SR. Pharmaceutical innovation by the seven UK-owned pharmaceutical companies (1964–1985). Br J Clin Pharmacol. 1988;25(3):387–96.

Chapter 3
Drug Absorption and Bioavailability

Debra Si Mui Sim

Abstract Most drugs are prescribed as oral preparations or extravascular injections (other than intravenous injections) for the treatment of systemic diseases. These drugs must therefore be absorbed in order to be transported to the target tissues to produce their pharmacological actions. Consequently, absorption plays a key role in determining whether or not a drug produces a clinical effect and how fast it occurs. The rate and extent to which a drug is absorbed systemically are related to its time-to-peak concentration (T_{max}) and fractional bioavailability (F). Often the two pharmacokinetic terms, absorption and bioavailability, are considered synonymously, but there is actually a subtle difference between them. It is possible for drugs to be well absorbed orally because of good lipid solubility and yet not have a good oral bioavailability because of extensive presystemic loss. While the intravenous bioavailability of drugs is always 100 %, the oral bioavailability is usually less than 100 % because of incomplete absorption and/or first-pass elimination. Many factors influence the oral bioavailability of a drug: some are related to the drug while others to the patient. To overcome poor bioavailability, we can increase the dose administered, change the pharmaceutical formulation, or use a different route of administration.

Keywords Drug absorption • Fractional bioavailability • Time-to-peak concentration • Ion trapping • Presystemic metabolism • First-pass effect

Introduction

Most drugs are prescribed as oral preparations or extravascular injections for the treatment of systemic diseases. These drugs must therefore be absorbed and delivered to the systemic circulation in order to be transported to the target tissues to produce their pharmacological actions. Consequently, absorption plays a key role

D.S.M. Sim, B.Sc., Ph.D. (✉)
Department of Pharmacology, Faculty of Medicine, University of Malaya,
50603 Kuala Lumpur, Malaysia
e-mail: debrasim@ummc.edu.my

© Springer International Publishing Switzerland 2015   17
Y.K. Chan et al. (eds.), *Pharmacological Basis of Acute Care*,
DOI 10.1007/978-3-319-10386-0_3

in determining whether or not a drug produces a clinical effect and how fast it occurs. The rate and extent to which a drug is absorbed systemically are related to its time-to-peak concentration (T_{max}) and fractional bioavailability (F).

Absorption or Bioavailability

Absorption refers to the transfer of a substance from its site of administration into the bloodstream. This means of course that intravenous route circumvents the absorption process whereas for all other routes, absorption is required for a drug to enter the body and produce its pharmacological effects. Bioavailability however is defined both as the fraction (F) of an administered dose of a substance that enters the systemic circulation in the unchanged form and the rate (T_{max}) at which it appears in the systemic circulation. Often these two pharmacokinetic terms (absorption and bioavailability) are considered synonymously, but there is actually a subtle difference between them. It is possible for drugs, such as morphine and nitroglycerin, to be well absorbed orally, as they are lipid soluble and can thus permeate intestinal mucosa readily and yet not have a good oral bioavailability because of extensive presystemic loss. This is because after crossing the intestinal wall, the absorbed drug enters the portal circulation and is carried to the liver before it appears in the systemic circulation (Fig. 3.1). This may result in a loss of drug through metabolism (especially by the cytochrome P450 enzymes) on its first passage through the liver before entering the systemic circulation. Hence, presystemic metabolism is also known as "first-pass effect".

Measurement of Bioavailability

For intravenous injection, the whole of drug dose is delivered directly into the systemic circulation, thus it will always have 100 % bioavailability (F = 1) and the maximal plasma concentration (C_{max}) is reached instantaneously ($T_{max} = 0$ min). For other parenteral routes such as subcutaneous and intramuscular injections, the bioavailability may still be close to 100 % for most therapeutic drugs (F = 0.75–1), since little or no significant metabolism of the drug occurs in the skin or muscle and these small molecules (MW <800) would have no problem permeating the capillary endothelium, but the time to reach its maximum plasma concentration may be relatively slower ($T_{max} > 0$ min) compared to the intravenous route. As for orally administered drugs, their bioavailabilities are often below 100 % (F < 1) because of incomplete absorption and/or first-pass elimination. Their C_{max} and T_{max} may also differ for the different dosage forms (Fig. 3.2).

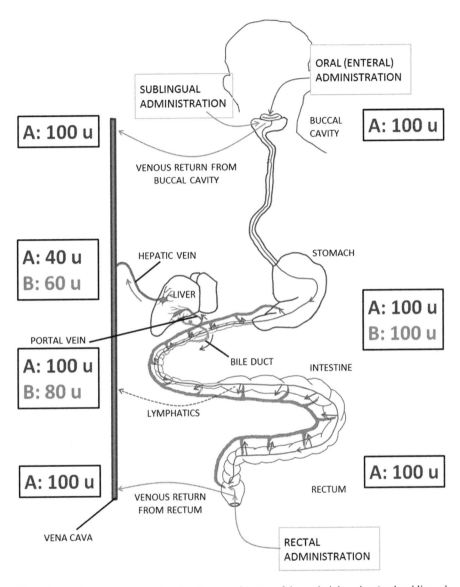

Fig. 3.1 A schematic diagram showing the enteral routes of drug administration (oral, sublingual and rectal) and their relative susceptibility to first-pass elimination, influencing the absorption and bioavailability of a drug. Following oral administration, Drug A is 100 % absorbed but 40 % bioavailable, while Drug B is 80 % absorbed and 60 % bioavailable. Hepatic first-pass elimination (1 – Fractional bioavailability) of Drug A is 60 % while that of Drug B is 40 %. Following sublingual administration, Drug A is 100 % bioavailable while following rectal administration, Drug A bioavailability is between 40 and 100 %, depending how far up the rectum the drug product is inserted

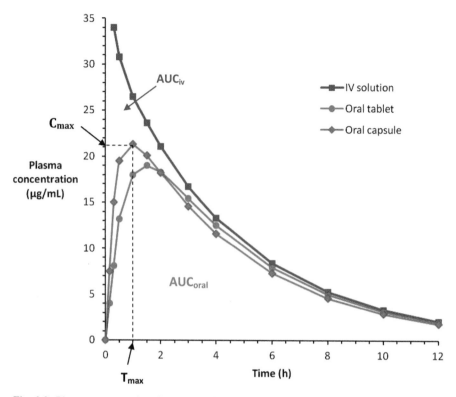

Fig. 3.2 Plasma concentration-time curves for a drug given by intravenous (IV) injection and two oral dosage forms (tablet and capsule). C_{max} = Peak concentration; T_{max} = Time-to-peak concentration; the oral bioavailability is determined by $F = AUC_{oral}/AUC_{iv}$, where AUC = area under the plasma concentration-time curve for the respective routes

---

Clinical Implication – Bioavailability of Parenteral Toxins

It is worth noting that in envenomation, the venom toxins (mostly large molecules such as peptides or proteins) that are delivered intramuscularly or subcutaneously may bind avidly to some tissue components or become destroyed in the surrounding of the bite sites before these toxins enter the bloodstream, resulting in incomplete bioavailability ($F < 1$). These initially unabsorbed toxins may form a reservoir at the bite site and later be released into the bloodstream causing relapse. This may be of significance during antivenom immunotherapy, especially if the antivenom plasma half-life is short. The patient is then exposed to the risk of the effects of the venom whilst the care provider makes the assumption that all the effects of the venom have been reversed.

Clinical Relevance of Bioavailability

A low systemic bioavailability will obviously result in the amount of drugs reaching the target tissue being reduced yielding a smaller than expected drug response. If the bioavailability is so poor, it is possible that the administered drug dose may not even reach the minimal effective concentration to produce the desired clinical effect. To overcome this, we can increase the dose administered, change the pharmaceutical formulation, or use a different route of administration.

Drug Permeation

During the process of absorption a drug must cross biological barriers in order to get from the site of administration into bloodstream. There are four major mechanisms by which drug molecules permeate cell membranes: diffusion through lipid, diffusion through aqueous channels, carrier-mediated transports (both active transport and facilitated diffusion) and pinocytosis. Of these mechanisms, diffusion through lipid and carrier-mediated transport are the ones most commonly encountered in drug absorption. The aqueous channels (about 0.4 nm in diameter) are too small to allow most drug molecules (usually >1 nm diameter) to pass through. Pinocytosis appears to be important for the transport of some big molecules (e.g., insulin across the blood brain barrier), but not for small molecules like most drugs.

Diffusion Through Lipid

Non-polar drug molecules passively diffuse across the membrane lipid according to its permeability coefficient, P, and the concentration difference across the membrane. Two physicochemical factors contribute to P, namely the solubility of the drug in the membrane lipid and its diffusivity. Lipid diffusion is by far the most important mechanism by which drugs cross intestinal mucosa to enter portal circulation.

Carrier-Mediated Transport

A number of lipid-insoluble drugs (e.g., levodopa, fluorouracil, iron and calcium) resemble endogenous substances and are carried across cell membranes by forming complexes with specific transmembrane proteins called carriers or transporters. This carrier-mediated transport may operate purely passively (as in facilitated diffusion) or it may be coupled to the electrochemical gradient of Na^+ (as in active transport). Carriers are proteins and like receptors, they exhibit selectivity and saturability, and are also subject to competitive inhibition. In addition, active transport may be blocked by inhibitors of cellular metabolism.

Factors Affecting Drug Absorption and Bioavailability

Oral Route

In most instances, absorption occurs when drug molecules are in the form of solutes. If a drug is given in solid form, then the drug must first break down into smaller particles (disintegration) and dissolve in the medium in which it is administered (dissolution) before it can traverse the cell membrane (permeation) and enter the bloodstream (absorption), Fig. 3.3. Thus the rate at which a given medication is absorbed will depend on the relative speed at which these processes occur, and the overall rate of absorption is often determined by the step with the slowest rate. Except for controlled-release medication, disintegration of a solid drug product usually occurs more rapidly than drug dissolution and permeation. Thus, in general, drug absorption may be dissolution rate-limited or permeation rate-limited.

Dosage Form and Formulation

Dosage form is essentially the pharmaceutical product in the form packaged for use (see Chap. 9). When the same drug is given in different dosage forms, its relative absorption rate will be in this descending order: solution/syrup > suspension > powder > capsule > tablet > coated tablet. This is because drugs in solution form would have avoided the steps of disintegration and

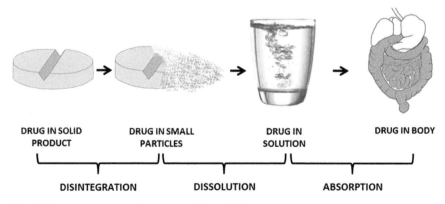

Fig. 3.3 Rate processes involved in drug bioavailability

dissolution. In contrast, coated tablets (e.g., enteric-coated erythromycin base) or special formulations (e.g., erythromycin stearate ester) are often designed to delay the disintegration and dissolution processes until the drug reaches the small intestine where the condition may be more favorable for its absorption.

The excipient used in different formulation can affect the bioavailability of a drug by influencing the rate of disintegration or dissolution and hence the liberation of the drug from the pharmaceutical preparation (e.g., phenytoin given with lactose as the excipient is better bioavailable compared to when given with calcium sulfate, resulting in unexpected toxicity when the former is replaced with the latter formulation).

Solubility of the Drug

Lipid solubility of a drug is often measured in terms of its octanol-to-water partition coefficient while the degree of ionization is related to its pK_a, acid dissociation constant, and the pH of the surrounding medium. For very polar and hydrophilic drugs (e.g., gentamicin and d-tubocurarine), dissolution in the aqueous intestinal medium is not a problem, but permeability through the lipid-rich cell membranes is and this becomes the rate-limiting step to the whole absorption process. The lipid solubility of these drugs is so poor that they are not effective when given orally.

However for a drug with very poor aqueous solubility such as ketoconazole, dissolution is the rate-limiting step in its absorption. Since ketoconazole is weakly basic, it dissolves better in the acidic environment of the gastric juice and absorption occurs more readily there than in the relatively more alkaline intestinal environment. Impairment of ketoconazole absorption secondary to achlorhydria (e.g., resulting from treatment with proton pump inhibitors or H_2 antagonists) has been documented for both healthy patients and patients with AIDS, while acidic beverages such as Coca-Cola enhances its absorption. Griseofulvin is another drug with very poor aqueous solubility and moderate lipid solubility and its absorption is increased with fatty meal.

Degree of Ionization

As most drugs are weak acids or bases, their degrees of ionization vary according to the pH at different levels of the gastrointestinal tract. Weak acids such as aspirin are less ionized at the low pH (1–3) of the stomach content and can easily diffuse cross the gastric mucosal membrane to reach the bloodstream on the serosal side. As the pH of the plasma is relatively more basic (pH = 7.4), this results in a relatively greater degree of ionization of the weak acid drug, which discourages back diffusion of the drug into the gastrointestinal lumen (Fig. 3.4). This unequal distribution of drug molecules based on the pH gradient across the gastric membrane is an example of "ion trapping". The reverse can occur with weak bases.

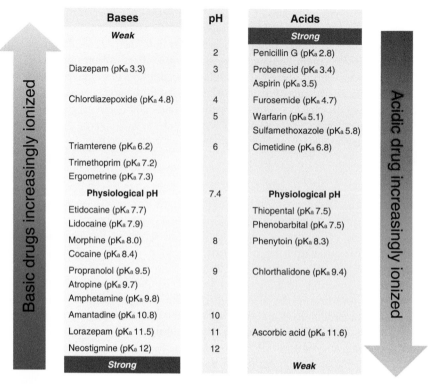

Fig. 3.4 The influence of pK_a on the degree of functional group ionization in basic and acidic drugs relative to the physiologic pH. The increasing depth of color in the arrows reflects increasing ionization relative to the physiologic pH of 7.4. Thus, for basic drugs, the more acidic the solution (i.e., the lower the pH), the greater is the proportion of drug ionized. The converse is true of the acidic drugs. [The pK_a of a drug is the pH at which 50 % of the drug molecules in solution are ionized]

The ion trapping of basic compounds within the gastric lumen is sometimes useful in forensic medicine. Drugs of abuse such as heroin, cocaine and amphetamine, are organic bases. Even when injected intravenously, they tend to accumulate in the stomach by crossing the gastric mucosa in the reverse direction. This drug behavior may find application in forensic medicine.

Mucosal Surface Area

Despite what is said about the effect of pH on the degree of ionization of a weak acid drug and hence its greater lipid solubility in the gastric lumen, the small intestine is still the optimal site of absorption due to the greatly increased mucosal surface area for absorption. The surface-to-volume ratio in the small intestine is so great that drugs ionized even to the extent of 99 % may still be effectively absorbed.

First-Pass Effect

This is also known as presystemic elimination, which may be caused by gut secretion (e.g., insulin is destroyed by digestive enzymes), or enzymes produced by gut microflora (e.g., digoxin is degraded by bacterial enzymes in the stomach), gut mucosa (e.g., epinephrine and levodopa are degraded by intestinal monoamine oxidase) and liver (e.g., lidocaine, morphine and nitroglycerin). Strictly speaking, hepatic first-pass metabolism does not affect oral absorption but it can affect oral bioavailability. Some drugs (e.g., tetracycline) form complexes with cations such as Ca^{2+}, Mg^{2+} and Fe^{2+} in the food or diet supplements and this may also result in presystemic loss. Oral absorption of alendronate is very poor (about 0.6 %) to start with and food decreases it even further. It is thus crucial that this drug should be taken on an empty stomach with just water.

Gut Motility

As the small intestine is the optimal absorption site for most drugs, a decrease in gastric emptying rate (e.g., migraine attack, fatty meals, antimuscarinic medication) generally delays their oral absorption although it may not significantly affect the extent of absorption unless the drug is not stable in the stomach. On the other hand, increased intestinal motility with prokinetic drugs such as metoclopramide would enhance the intestinal absorption of most drugs. Excessively enhanced gut motility (e.g., during diarrhea) however may reduce oral absorption of drugs with poor lipid solubility due to insufficient time for drugs to cross the intestinal mucosa.

Splanchnic Blood Flow

Physiological changes in blood flow have little impact on the rate of extent of gastrointestinal absorption of most drugs. However, splanchnic blood flow may be greatly reduced in shock to such an extent that it may slow the absorption of some drugs.

Non-oral Routes

For other routes of drug administration, the absorption depends mainly on regional blood flow, area of absorption surface, drug properties and formulation in much the same way as are described for oral absorption.

Key Concepts

- Absorption is the process of transferring a substance from its site of administration into the bloodstream.
- Bioavailability is defined both as the fraction (F) of an administered dose of a substance that enters the systemic circulation in the unchanged form and the rate (T_{max}) at which it appears in the systemic circulation.
- A low systemic bioavailability can be overcome by increasing the dose administered, changing the pharmaceutical formulation, or using a different route of administration.
- Most drugs are well absorbed from the gut, and lipid diffusion is the most important mechanism for oral absorption.
- Absorption from the gut depends on many factors, some of which are related to the drugs (e.g., lipid solubility, degree of ionization) and others to the patients (e.g., gut motility, intestinal pH, splanchnic blood flow).

Summary

Drugs that are prescribed as oral preparations or extravascular injections must enter systemic circulation in order to be able to produce a systemic effect. Consequently, absorption plays a key role in determining whether or not a drug produces a clinical effect and how fast it occurs. The rate and extent to which a drug is absorbed systemically are related to its time-to-peak concentration (T_{max}) and fractional bioavailability (F). A drug may be well absorbed orally because of good lipid solubility and yet not has a good oral bioavailability because of extensive presystemic loss. While the intravenous bioavailability of drugs is always 100 %, the oral bioavailability is usually less than 100 % because of incomplete absorption and/or first-pass elimination. Many factors influence the oral bioavailability of a drug; some are related to the drug while others to the patient. To overcome poor bioavailability, we can increase the dose administered, change the pharmaceutical formulation, or use a different route of administration.

Further Reading

1. Boucher BA, Wood GC, Swanson JM. Pharmacokinetic changes in critical illness. Crit Care Clin. 2006;22:255–71.
2. Chin TWF, Loeb M, Fong IW. Effects of an acidic beverage (Coca-Cola) on absorption of ketoconazole. Antimicrob Agents Chemother. 1995;39(5):1671–5.
3. Hughes SG. Prescribing for the elderly patient: why do we need to exercise caution? Br J Clin Pharmacol. 1998;46:531–3.
4. Rang HP, Dale MM, Ritter JM, Flower RJ, Henderson G. Rang and Dale's pharmacology. 7th ed. Edinburgh: Churchill Livingstone; 2012.

Chapter 4
Drug Distribution

Debra Si Mui Sim

Abstract Once a drug enters the bloodstream, it will be carried by the blood to various parts of the body. In order for it to act on its target site(s) of action, the drug must leave the bloodstream to which it may later return. Such reversible transfer of substances between the blood and extravascular tissues is known as distribution. Distribution generally occurs rapidly for most drugs and is often much faster than elimination. How widespread a drug action is often depends on its distribution profile. Its ability to distribute to specific tissues depends on both physiological factors (e.g., tissue perfusion, membrane permeability) and drug properties (e.g., molecular size, degree of ionization, lipid solubility, relative binding to plasma protein and tissue protein). The volume of distribution (V_d) is the second most important pharmacokinetic parameter after plasma clearance (CL). It gives some idea as to the extent of distribution of a drug in the body. It also determines the loading dose (D_L) to be given during multiple dosing in order to achieve plasma steady-state concentration (C_{ss}) faster. Factors that alter the distribution of a drug (e.g., edema, sepsis, pregnancy) may contribute to failure in achieving the expected clinical outcome.

Keywords Volume of distribution • Plasma protein binding • Tissue protein binding • Perfusion-limited • Permeability-limited

Introduction

Once a drug enters the bloodstream, it will be carried by the blood to various parts of the body. In order for it to act on its target site(s) of action, the drug must leave the bloodstream to which it may later return. Such reversible transfer of substances between the blood and extravascular tissues is known as distribution. Distribution generally occurs rapidly for most drugs and is often much faster than elimination.

D.S.M. Sim, B.Sc., Ph.D. (✉)
Department of Pharmacology, Faculty of Medicine, University of Malaya,
50603 Kuala Lumpur, Malaysia
e-mail: debrasim@ummc.edu.my

© Springer International Publishing Switzerland 2015
Y.K. Chan et al. (eds.), *Pharmacological Basis of Acute Care*,
DOI 10.1007/978-3-319-10386-0_4

How widespread a drug action is often depends on its distribution profile. Its ability to distribute to specific tissues depends on both physiological factors and drug properties.

How Drugs Are Distributed

Drugs are transported to different parts of the body mainly by the circulatory system. The transfer of drugs between the blood and tissues takes place largely in the capillary bed. The capillary wall forms the blood-tissue barrier across which drugs must permeate to enter into the interstitial fluid. This is followed by the permeation of drugs from the interstitial fluid (ISF) to the intracellular fluid (ICF) through the membrane of the tissue cell (Fig. 4.1). As most drug receptors are located on the surface of cell membranes, it is not always necessary for drugs to enter the cells in order for it to be effective.

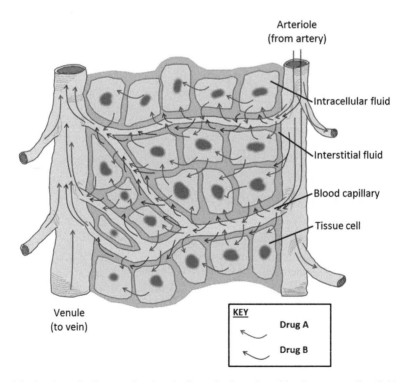

Fig. 4.1 A schematic diagram showing the flow of solutes from blood to surrounding fluids and tissue-cells. Drug A, being more lipid soluble than Drug B, enters both interstitial (ISF) and intracellular fluids (ICF), whereas the polar Drug B, is only able to cross over into the interstitial fluid

How Do Drugs Traverse Blood-Tissue Barriers

The ease with which drugs cross blood-tissue barrier depends on the structural and functional characteristics of the capillary endothelial cells at those tissue sites (Fig. 4.2). In most of the capillary beds, such as those of muscle, fat and nervous tissues, the capillary endothelium lacks pores and the cells are joined by tight non-permeable junctions. In other capillary beds, such as those of heart muscle, the endothelial cell allows the transport of fluid or macromolecules (e.g., insulin) in vesicles from the blood into the interstitium and vice versa. In some capillary beds, such as in the pancreas, gut, kidney glomeruli and endocrine glands, the endothelial lining behaves as though it were fenestrated by pores. In the liver, spleen and bone marrow, open spaces exist between endothelial cells, called intercellular clefts,

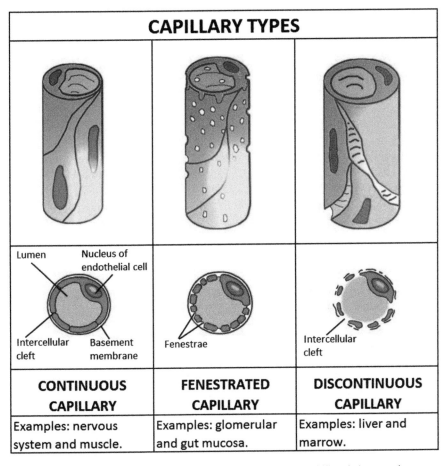

Fig. 4.2 Different types of capillaries with varying degrees of permeability, their respective cross-sections and examples of typical locations where these capillary types are found

which allow free exchange of drugs (most of which have molecular weights of less than 900) between blood and the interstitium.

In summary, the capillary wall basically consists of an endothelial cell layer and a basement membrane enveloping it. The permeation of drugs across this capillary wall can occur by one of several processes such as passive diffusion, transcytosis (vesicular transport) and bulk flow as well as by carrier-mediated transport mechanisms such as facilitated diffusion or active transport.

Factors Affecting the Rate of Drug Distribution

After a drug enters the bloodstream, its rate of distribution to the various body tissues is governed predominantly by two factors, each of which is potentially rate-limiting: (1) the rate of blood flow (perfusion) to the tissue and (2) the ability of the drug to cross membranes (permeability) to enter the tissue.

For drugs which have high membrane permeability (e.g., thiopental, steroids and benzodiazepines), the rate of distribution to tissues depends mainly on the rate of blood flow to these tissues. Thus, tissue drug concentration would increase most rapidly in highly perfused tissues such as kidneys, liver, lung, heart and brain, more slowly in less well perfused tissues such as skeletal muscle, but very slowly in poorly perfused tissues, such as bone, fat and skin. The distribution of these lipophilic drugs is described as perfusion-limited.

On the other hand, drugs with poor membrane permeability (e.g., penicillins, aminoglycoside antibiotics, succinylcholine and vecuronium) would distribute more readily to tissues such as bone marrows, spleen, liver and muscle, where the capillary wall is more "leaky", than to the brain, where the capillary wall seems to be impermeable. This selective distribution occurs despite a large amount of the drug being delivered to the highly perfused brain, and is an example of permeability-limited drug distribution.

Clinical Correlation – Distribution of Perfusion-Limited Versus Permeability-Limited Drugs

This means therefore lipophilic drugs such as thiopental gets into the brain more quickly than into skeletal muscle, whereas polar drugs such as vecuronium gets into skeletal muscle faster than into the brain. It also means that when there is a marked change in tissue perfusion, such as during circulatory shock, the distribution of drugs which are perfusion-limited would be affected more than those which are permeability-limited. Conversely, when there is a marked change to the tissue pH or amount of proteins in plasma, as may occur in sepsis, severe burns or traumatic injuries, the distribution of permeability-limited drugs to those areas would be affected more than that of perfusion-limited drugs.

Factors Affecting the Extent of Drug Distribution

Distribution is a reversible process. Hence, the exchange of drug molecules across a barrier would continue until the concentrations of the free drugs on both sides of the barrier are the same. While tissue perfusion has a major effect on the rate of distribution of drugs to various tissues, drug properties and membrane permeability play a greater role on the extent of drug distribution at equilibrium.

Important physicochemical properties of drugs which affect their extent of distribution include molecular size, degree of ionization, octanol:water partition coefficient and their relative affinities for tissue and plasma proteins.

Drugs with large molecular weight (e.g., heparin) and those which are extensively bound to plasma protein (e.g., warfarin) have difficulty crossing the capillary wall into the interstitium. Their distribution would be restricted to mainly the plasma compartment (3 L/70 kg). Polar drugs such as gentamicin and vecuronium cross the capillary endothelium readily (by bulk flow) into the extravascular fluid compartment (14 L/70 kg) but cannot diffuse into the tissue cell. In contrast, lipophilic drugs such as diazepam and theophylline diffuse easily into intracellular fluid compartment and distribute into total body water (42 L/70 kg). Some drugs (e.g., digoxin and chloroquine) bind extensively to extravascular tissue proteins resulting in volume of distribution (V_d) values that far exceed the total body water volume. This phenomenon can be observed even with drugs (e.g., fluoxetine and imipramine) which have extensive plasma protein binding (Table 4.1).

The volume of distribution (V_d) is the second most important pharmacokinetic parameter after plasma clearance (CL). It determines the loading dose (D_L) to be given during multiple dosing in order to achieve plasma steady-state concentration (C_{ss}) faster, according to the equation $D_L = V_d C_{ss}$. An obvious consequence of having a large V_d is that a large loading dose is needed for this purpose.

Ageing is accompanied by changes in body fat, lean body mass and total body water. These changes result in reduced V_d of water soluble drugs, e.g., digoxin, (which may lead to increased initial drug concentration) and increased V_d of lipophilic drugs, e.g., benzodiazepines (which may lead to increased elimination half-life and prolonged effect). Both types of drug may therefore require a reduction in dose and/or dose interval.

Table 4.1 Volumes of distribution (V_d) and degree of plasma protein binding for selected drugs

Drug	Bound in plasma (%)	V_d (L/70 kg)
Heparin	Extensive	4
Warfarin	99	10
Gentamicin	<10	22
Vecuronium	69	25
Theophylline	56	35
Diazepam	99	77
Digoxin	25	500
Chloroquine	61	13,000
Fluoxetine	94	2,450
Imipramine	90	1,300

Binding of Drugs to Plasma Proteins

In the plasma, drugs may bind to various proteins, primarily albumin but also to α_1-acid glycoprotein, globulins (low capacity but high affinity) and lipoproteins. This binding is reversible and occurs to varying extents for different drugs (e.g. <10 % for gentamicin to ≥ 99 % for warfarin and diazepam). Many acidic drugs, such as salicylates and penicillins, and some neutral or basic drugs bind to albumin (low affinity but high capacity), while α_1-acid glycoprotein binds mainly basic drugs such as lignocaine and propranolol. Plasma globulins (α, β, γ-globulins) have low capacity but high affinity for the binding of endogenous substances such as corticosteroids.

Drugs exist in the plasma in the unbound (i.e. free) and bound (to plasma proteins) forms, which are in equilibrium with each other.

$$Drug \; + \; Protein \leftrightarrow Drug\text{-}Protein$$

The protein-bound drug is a large complex which cannot readily permeate cell membranes to enter the target site; hence restricting its distribution and also its pharmacological activity. Furthermore, protein-bound drugs are also not excreted readily.

Plasma protein binding of drugs is subject to saturation and competitive displacement. This is because there are limited binding sites on the protein for binding drugs at any one time. Therefore, drugs with a higher affinity for the same binding site on the protein would displace the bound drugs, resulting in a transient increase in the unbound concentration of the displaced drug. This could potentially lead to enhanced drug actions. However, drug-drug interactions involving protein-binding displacement are generally not clinically important as the displaced drug in the body would be eliminated readily by the liver or kidneys, thus bringing down the concentration of the unbound drugs.

Selective Accumulation of Drugs in Tissues

There are various ways by which drugs may accumulate selectively in certain tissues. Below are some examples of selective drug accumulation.

1. Binding to specific tissue components

The broad spectrum antibiotic tetracycline binds with cations such as calcium to form tetracycline-calcium complexes that irreversibly deposit in the bones and teeth of developing fetus or young children (below 8 years old). This results in the retardation of bone growth and discoloration of teeth.

2. Presence of active transport

Following ingestion (accidental or intentional), the herbicide paraquat is actively taken up into the lungs from the plasma by a selective transport system meant for polyamines. Consequently, the herbicide selectively accumulates in the lungs where it may attain concentration which is several times higher than that in the plasma, even when the plasma concentration is falling. This selective distribution results in organ-specific toxicity (lung fibrosis) typical of paraquat poisoning.

3. Very high lipid solubility

The intravenous anesthetic thiopental is highly lipid soluble and passes into the brain easily. Since the brain is highly perfused, the drug reaches its target site very rapidly and produces a very fast induction of anesthesia. However, as the drug is progressively taken up by the less well perfused tissues such as muscle and then eventually by the poorly perfused adipose tissues, it results in a decrease in the plasma drug concentration. The change in the concentration gradient between plasma and brain now favors the back diffusion of the anesthetic drug from the brain into the plasma and then to other tissues. This redistribution of thiopental from brain to other less well perfused tissues is mainly responsible for the rapid termination and short duration of its anesthetic action.

4. Ion-trapping effect

Trimethoprim is a lipophilic weak base ($pK_a = 7.3$) and readily diffuses across cell membranes. Once it enters the prostate gland it would become more ionized in the relatively more acidic prostatic fluid (pH ~6.5) environment, compared to the pH 7.4 of plasma. This prevents the antibiotic from diffusing back into plasma. Such ion-trapping effect helps to concentrate the antibiotic at the target site and contributes to its effectiveness in treating bacterial prostatitis.

Physiological Barriers to Drug Distribution

Besides the simple capillary endothelial barrier mentioned above, there are some more specialized physiological barriers which restrict the permeability of drugs into specific tissues.

1. Blood–brain barrier

In many tissues, the capillary wall forms a continuous sheet with no intercellular clefts or intracellular pores (Fig. 4.2). In the brain, the capillary endothelial cells are joined by tight junctions and pericapillary glial cells are present, making this blood-tissue barrier impermeable to many drugs, including many antineoplastic drugs given for treating brain tumor or antibiotics for treating central nervous system infection. In addition, efflux carriers like P-glycoproteins and organic anion-transporting polypeptides (OATP) transport drugs out of the brain and other tissues which express these transporters. However, the permeability of this blood–brain barrier increases somewhat during inflammation (e.g., meningitis) and it allows sufficient amount of penicillin to enter to produce its bactericidal effect.

Although dopamine cannot permeate blood–brain barrier due to poor lipid solubility, its precursor levodopa is able to do so because of the presence of amino acid transporters which can also carry levodopa into the brain. Once in the brain, levodopa is converted to its active drug dopamine. Some enzymes (e.g., monoamine oxidase) are also present in the capillary walls of the brain and these would destroy drugs such as catecholamine (e.g., epinephrine and norepinephrine) from entering the brain.

Clinical Correlation-Importance of "Continuous" Capillaries (Fig. 4.2) in the brain

The brain is impermeable to the catecholamines as these would have been destroyed by enzymes such as monoamine oxidase present in the capillary walls. This prevents the blood vessel of the brain from vasoconstriction and compromising oxygen delivery each time we administer these drugs or when it is produced endogenously by the adrenals in times of stress.

---

2. Placental barrier

The maternal blood supply is separated from the fetal blood supply by a layer of trophoblast cells which together constitute the so called placental barrier. It is worth noting that drug molecules permeate the placental barrier more easily than the blood–brain barrier. Restricted amounts of water soluble drugs, especially if present in high concentrations for a long period, may cross the placenta into the fetal circulation. Thus, the use of drugs during pregnancy must be with great caution as many of these drugs may cause harm to the fetus (i.e., teratogenic). For example, ingestion of thalidomide during pregnancy causes phocomelia in the newborn while maternal consumption of alcohol can lead to fetal alcohol syndrome.

Volume of Drug Distribution

It is practically impossible to measure the exact extent of distribution of a drug in the body. We can however estimate the extent of drug distribution mathematically assuming the body is a homogeneous compartment into which a drug distributes. Therefore the volume of distribution (V_d) is the volume into which a drug appears to have distributed assuming the tissue concentrations equal to that of plasma. It is also a proportionality constant that relates the amount of drug in the body (A_b) to its plasma concentration (C_p) at any one time.

$$A_b = V_d C_p$$

Table 4.2 How the distribution and elimination of drugs affects their plasma half-lives

Pharmacokinetic parameters	Dirithromycin (macrolide antibiotic)	Tenoxicam (NSAID)
V_d (L)	800	9.6
CL (L/h)	38.0	0.106
$t_{1/2}$ (h)	44	67

Relationship Between Drug Distribution and Drug Elimination Half-life

Although drug distribution and elimination are described as two separate pharmacokinetic processes, and V_d and CL are two independent pharmacokinetic parameters, yet the extent of drug distribution (whether extensive or restrictive) can have an effect on the elimination half-life of a drug, according to the equation $t_{1/2} = 0.693 V_d / CL$.

In the examples given in Table 4.2, a long elimination half-life can occur with both a drug with large V_d (e.g., dirithromycin) as well as with a drug with small V_d (e.g., tenoxicam). In the case of dirithromycin, the long elimination half-life is a result of the extensive distribution of the drug in tissues, despite its large total body clearance (CL). In contrast, the long elimination half-life of tenoxicam results more from restrictive drug clearance due to the binding of the drug to plasma protein, making it hard for the drug to be cleared rapidly, despite its small volume of distribution (V_d).

Key Concepts

- The ability of a drug to distribute to specific tissues depends on both physiological factors (e.g., tissue perfusion, membrane permeability) and drug properties (e.g., molecular size, degree of ionization, lipid solubility, relative binding to plasma protein and tissue protein).
- Volume of distribution (V_d) is defined as the volume into which a drug appears to have distributed assuming the tissue concentrations equal to that of plasma.
- Polar drugs are confined to plasma and interstitial fluids (generally small V_d) and most do not enter the brain following acute dosing. Their distribution is generally permeability-limited.
- Lipid-soluble drugs reach all fluid compartments (generally large V_d) and may accumulate in adipose tissue with repeated dosing. Their distribution is generally perfusion-limited.
- For drugs that accumulate outside the plasma compartment, their V_d may exceed total body volume.
- Both changes in V_d and CL can influence the plasma half-life of a drug.

Summary

Distribution is the reversible transfer of substances between the blood and extravascular tissues. The ability of a drug to distribute to specific tissues depends on both physiological factors (e.g., tissue perfusion, membrane permeability) and drug properties (e.g., molecular size, degree of ionization, lipid solubility, relative binding to plasma protein and tissue protein). The volume of distribution (V_d) is the second most important pharmacokinetic parameter after plasma clearance (CL). It gives some idea as to the extent of distribution of a drug in the body. It also determines the loading dose (D_L) to be given during multiple dosing in order to achieve plasma steady-state concentration (C_{ss}) faster. Factors that alter the distribution of a drug (e.g., edema, sepsis, pregnancy) may contribute to failure in achieving the expected clinical outcome.

Further Reading

1. Begg EJ. Instant clinical pharmacology. Malden, Massachusetts: Blackwell; 2003.
2. Boucher BA, Wood GC, Swanson JM. Pharmacokinetic changes in critical illness. Crit Care Clin. 2006;22:255–71.
3. Brunton LL, Lazo JS, Parker KL, editors. Goodman & Gilman's The pharmacological basis of therapeutics. 11th ed. New York: McGraw-Hill; 2006.
4. Hughes SG. Prescribing for the elderly patient: why do we need to exercise caution? Br J Clin Pharmacol. 1998;46:531–3.
5. Shargel L, Yu ABC. Applied biopharmaceutics and pharmacokinetics. 4th ed. New York: McGraw Hill; 1999.

Chapter 5
Drug Elimination

Debra Si Mui Sim

Abstract Most drugs, especially toxicants, are lipophilic in nature and once absorbed, would remain inside the body, distributing between tissues indefinitely unless they are converted to polar substances and excreted from the body. Drug elimination is the process that permanently removes drugs from the body. It may occur physically by excretion or chemically by metabolism, and is responsible for the termination of most drug actions. Drug metabolism is the single most important pharmacokinetic process that accounts for many of the inter-individual variations seen in therapeutic drug responses, and the liver is the main organ involved. Both internal (e.g., age, diseases or genetic factors) and external (e.g., diet or environment) factors may affect the rate or extent of drug metabolism. The metabolism of drugs usually results in water soluble metabolites which are readily excreted, and this generally occurs in the kidneys. The processes involved in the renal excretion of drugs are glomerular filtration, active tubular secretion and passive tubular reabsorption. While the former two processes facilitate drug excretion and are not affected by urine pH, the latter one decreases drug excretion and is pH-dependent. The two pharmacokinetic parameters that are related to drug elimination are drug clearance (CL) and plasma half-life ($t_{1/2}$). Drugs with high clearance (e.g., lidocaine, propranolol, morphine) display flow-dependent elimination, whereas drugs with low clearance (e.g., phenytoin, ethanol) exhibit capacity-limited elimination.

Keywords Drug elimination • Biotransformation (metabolism) • Excretion • Plasma clearance • Elimination half-life

Introduction

Most drugs, especially toxicants, are lipophilic in nature and once absorbed, would remain inside the body, distributing between tissues indefinitely unless they are converted to polar substances and excreted from the body. Drug elimination is the process that permanently removes drugs from the body. This may occur physically

D.S.M. Sim, B.Sc., Ph.D. (✉)
Department of Pharmacology, Faculty of Medicine, University of Malaya,
50603 Kuala Lumpur, Malaysia
e-mail: debrasim@ummc.edu.my

© Springer International Publishing Switzerland 2015
Y.K. Chan et al. (eds.), *Pharmacological Basis of Acute Care*,
DOI 10.1007/978-3-319-10386-0_5

by excretion or chemically by metabolism and is responsible for the termination of most drug actions. Drug elimination especially metabolism, is also the single most important pharmacokinetic process that is responsible for many of the inter-individual variations seen in therapeutic drug responses. It is usually the slowest of the four major pharmacokinetic processes. Hence, the terminal plasma half-life of a drug usually describes its elimination half-life.

Drug Elimination by Metabolism or Biotransformation

Biotransformation and metabolism are often used synonymously in describing the disposition of a drug. However, biostransformation is a more accurate term as it describes the process by which a substance is chemically transformed into another entity by the body. On the other hand, the term metabolism is used frequently to describe also the total fate of a xenobiotic (drug or non-drug chemical) in the body. In pharmacokinetics, the term metabolism is commonly used to mean biotransformation from the standpoint that the products of drug biotransformation are called metabolites.

Sites of Drug Metabolism

Liver is the principal, but not the only, organ involved in metabolizing drugs since a large variety of enzymes reside there. Other important sites of drug metabolism include gastrointestinal tract (e.g., catecholamines), lungs (e.g., prostaglandins, angiotensins), kidneys (e.g., imipenem) and plasma (e.g., succinylcholine, procaine).

Process of Drug Metabolism

Drug metabolism involves chemical reactions (usually enzymatic) that produce metabolites that are frequently more polar, more readily excreted and often biologically inactive. But at times, active drugs (e.g., diazepam, aspirin, codeine) may be converted to active metabolites (e.g., oxazepam, salicylate, morphine, respectively), which will then prolong the pharmacological actions of the parent drugs. Drugs (e.g., acetaminophen, cyclophosphamide) may occasionally produce reactive metabolites (e.g., N-acetyl-p-benzoquinoneimine, acrolein, respectively) that may be toxic to intracellular components (see Fig. 27.2 in Chap. 27). Drug metabolism may also be used to convert an inactive precursor (called a prodrug) to the actual active drug in the body. Examples include:

(a) levodopa to dopamine (in the central nervous system)
(b) prednisone to prednisolone (in the liver)

(c) erythromycin succinate to erythromycin (in the gastrointestinal tract)

(d) zidovudine (AZT) to AZT-triphosphate (in infected cells)

Advantages of prodrugs include improved bioavailability, prolonged duration of action, and site-specific drug delivery.

Phases of Drug Metabolism

Drug metabolism reactions are broadly divided into two phases: Phase I (or functionalization) reactions and phase II (or conjugative) reactions. Examples of drugs undergoing the various phase I and II reactions are given in Table 5.1.

Phase I reactions add or unmask functional groups (e.g., $-OH$, $-NH_2$, $-SH$, $-COOH$, etc.), which can then participate in phase II reactions. These are non-synthetic reactions. Oxidation reactions occur mainly on the smooth endoplasmic reticulum (sER) and are catalyzed by a family of enzymes known as microsomal "mixed-function oxidases" (MFO, so called because they catalyze a large range of oxidation reactions on a variety of substrates). These microsomal oxidative enzymes are also known as "monooxygenases". Other phase I reactions (including some oxidation reactions) take place in the cytosol, plasma and mitochondria. Phase I metabolites are usually not much more polar than their parent drugs, but they are often chemically reactive and may even be toxic (e.g., acetaminophen-induced liver toxicity or cyclophosphamide-induced bladder toxicity).

Table 5.1 Examples of drugs undergoing phase I and II drug metabolism reactions

Phase I reactions	Drugs
Oxidation (involving cytochrome P450)	Amphetamine, codeine, diazepam, ethanol, lignocaine, phenobarbital, phenytoin
Oxidation (others)	Epinephrine, ethanol, theophylline, tyramine
Reduction	Chloramphenicol, halothane, methadone
Hydrolysis	Esters: procaine, succinylcholine
	Amides: lignocaine, procainamide

Phase II reactions	Drugs
Glucuronide conjugation	Acetaminophen, morphine
Sulfate conjugation	Acetaminophen, sex steroids
Glutathione conjugation	Acetaminophen
Glycine conjugation	Salicylates
Acetylation	Isoniazid
Methylation	Catecholamines

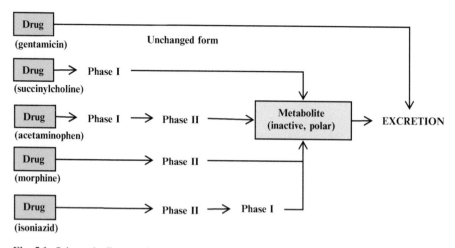

Fig. 5.1 Schematic diagram showing different pathways of drug metabolism

Phase II reactions are synthetic and involves the addition of a large endogenous substrate to the drug or its phase I metabolite to form a conjugate. Thus, phase II metabolites are frequently much more polar and almost always inactive, except for a few drug metabolites (e.g., morphine 6-glucuronide, N-acetylprocainamide or minoxidil sulfate). These conjugation reactions occur mainly in the cytosol, except for glucuronide conjugation reactions, which occur on sER and thus glucuronyl-transferases are microsomal enzymes. Generally speaking, only the microsomal enzymes are inducible.

Drugs may be excreted without being metabolized (e.g., gentamicin). However, the majority of drugs undergo some form of metabolism (whether phase I or phase II, or more commonly, both) before excretion. While most drugs undergo phase I and phase II reactions in that order, in some cases (e.g., isoniazid), phase II reactions precede phase I reactions (Fig. 5.1).

Cytochrome P450

Cytochrome P-450 (P450) is a heme-containing enzyme, which is part of the MFO system, and can exist in many isoforms (e.g., CYP1A2, CYP2D6, CYP3A4). These P450 enzymes are predominantly expressed in the liver and are extremely important in drug metabolism. They are essential for most drug metabolism, contribute to most of the observed variations in drug response of differing ethnic origins (due to genetic polymorphism of P450 isozymes), account for many of the important drug interactions (Table 5.2), and are responsible for a number of serious adverse effects. In view of this, physicians should be cautious in prescribing drugs known to be P450 inducers or inhibitors, especially to critically ill patients.

Table 5.2 Important P450 enzymes and examples of drugs which require dose adjustment to prevent adverse effects resulting from drug interaction involving enzyme inhibition or induction

Enzymes	Potent inhibitor	Potent inducer	Affected drug
CYP1A2	Amiodarone	Carbamazepine	Clozapine
	Cimetidine	Phenobarbital	Haloperidol
	Ciprofloxacin	Rifampin	Propranolol
	Fluvoxamine	Cigarette smoke	Theophylline
CYP2D6	Amiodarone	No significant	Amitriptyline
	Cimetidine	inducers	Carvedilol
	Diphenhydramine		Codeine
	Fluoxetine		Donepezil
	Quinidine		Haloperidol
	Ritonavir		Metoprolol
			Propranolol
			Risperidone
			Tramadol
CYP3A4	Clarithromycin	Carbamazepine	Alprazolam
	Cimetidine	Phenobarbital	Amiodarone
	Erythromycin	Phenytoin	Astemizole
	Fluconazole	Rifampin	Carbamazepine
	Indinavir	St. John's wort	Cyclosporine
	Omeprazole		Diazepam
	Sertraline		Fentanyl
			Simvastatin
			Zolpidem

Factors Affecting Drug Metabolism

Both internal (e.g., age, diseases or genetic factors) and external (e.g., diet or environment) factors may affect the rate or extent of drug metabolism.

1. **Age:** The activity of hepatic microsomal enzymes is low in both the very young and the elderly patients resulting in impaired drug metabolism. Hence there is an increased risk of drug toxicity with reduced inactivation of drugs. For example, the grey baby syndrome associated with the use of chloramphenicol in neonates is the result of an inadequate amount of glucuronyl transferases to conjugate the drug in the immature liver. On the other hand, phenobarbital, a potent enzyme inducer, can be given to accelerate the clearance of bilirubin in jaundiced neonates by enhancing bilirubin conjugation.

2. **Diseases:** Chronic liver diseases (e.g., cirrhosis) markedly affect the hepatic metabolism of certain drugs (e.g., diazepam) and prolong their duration of action in the body, leading to potential risk of overdose. Therefore, drugs which are highly dependent on liver for their elimination (e.g., opioids, acetaminophen, and propranolol) should be used with caution in patients with liver diseases and their dosages reduced accordingly.

3. **Genetic variation:** The metabolism of certain drugs (e.g., succinylcholine and isoniazid) is significantly reduced in susceptible people who have either a defective enzyme (e.g., atypical pseudocholinesterase for hydrolyzing succinylcholine) or abnormally low level of an enzyme (e.g., acetylation of isoniazid).
4. **Diet:** Certain plant foods (e.g., cabbage and cauliflower) and barbecued foods contain bioactive compounds which are enzyme inducers. Regular ingestion of such foods can significantly increase the metabolism of administered drugs and reduce their therapeutic effectiveness. On the other hand, grapefruit juice is known to inhibit certain P450 enzymes that metabolize a number of drugs and may result in increased risk of toxicity.
5. **Environment:** Cigarette smoke and dichlorodiphenyltrichloroethane (DDT, a banned insecticide in many countries) are also powerful inducers of drug-metabolizing enzymes.

Drug Elimination by Excretion

The metabolism of drugs usually results in water soluble metabolites which are readily excreted. Some drugs, however, are directly eliminated without metabolism. While the liver is the major organ for eliminating drugs by metabolism, it is the kidneys which are responsible for eliminating most drugs and metabolites by excretion. Besides the kidneys, the intestine, the biliary system and the lungs are also involved in excreting many drugs, while a small amount of drugs may be excreted in the saliva, sweat and milk.

Renal Excretion

Three processes are involved in the renal excretion of drugs (Fig. 5.2). They are glomerular filtration, active tubular secretion and passive tubular reabsorption. While the former two processes facilitate drug excretion and are not affected by urine pH, the latter one decreases drug excretion and is pH-dependent.

1. *Glomerular filtration*: Drugs with MW <5,000 are readily filtered. The concentration of drugs in the filtrate is directly proportional to the glomerular filtration rate (GFR) and the fraction of unbound drug in the plasma.
2. *Active tubular secretion*: Active secretion of acidic drugs (e.g., penicillin, salicylates, furosemide, probenecid, sulfonamides) and basic drugs (e.g., amphetamine, quinine, procaine, morphine) occur in the proximal renal tubules. These transport systems are saturable and subject to competition by drugs with similar physicochemical properties. Thus, probenecid may competitively inhibit the active secretion of penicillin, thereby prolonging the duration of penicillin action. On the other hand, competitive inhibition of the tubular secretion of methotrexate by probenecid and salicylates contributes to its increased risk of nephrotoxicity.

THE NEPHRON

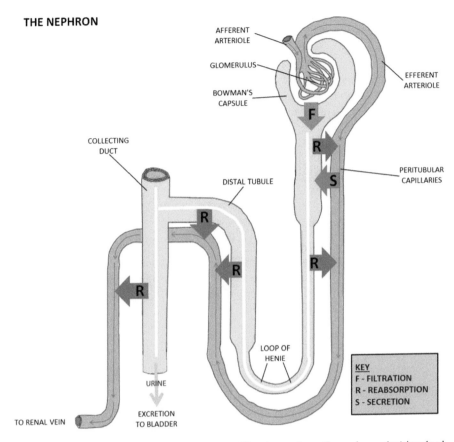

AFFERENT
ARTERIOLE

GLOMERULUS

EFFERENT
ARTERIOLE

BOWMAN'S
CAPSULE

F

COLLECTING
DUCT

R

PERITUBULAR
CAPILLARIES

DISTAL TUBULE

S

R

R

R

R

R

LOOP OF
HENIE

KEY
F - FILTRATION
R - REABSORPTION
S - SECRETION

URINE

TO RENAL VEIN

EXCRETION
TO BLADDER

Fig. 5.2 The nephron showing the three processes (filtration, reabsorption and secretion) involved in the excretion of drugs

3. *Passive tubular reabsorption*: This process is affected by urinary pH and the degree of ionization of a drug. Weakly acidic drugs such as aspirin and pheno-barbital are more ionized in alkaline urine and therefore more likely to be excreted because the ionized drugs are not reabsorbed from the tubule. The same is true with weakly basic drugs such as amphetamine and quinine in the presence of acidic urine. This property is used in the treatment of poisoning where forced alkaline diuresis is employed to hasten excretion of salicylates and barbiturates, whereas forced acid diuresis can be used to enhance the excretion of amphetamine and quinine.

Biliary and Fecal Excretion

There is a portion of orally administered drugs that are not absorbed and these are excreted in the feces. These are often large polar drugs. Large water-soluble conjugates produced in the liver are excreted via the bile into the intestine. Some of these conjugated metabolites may get broken down in the lower portion of the gastrointestinal tract to release more lipid soluble drugs, which are reabsorbed and carried back to the liver. This cycle is then repeated. Such recycling is known as "enterohepatic circulation" and is responsible for prolonging the duration of action of drugs such as oral contraceptive steroids, morphine and erythromycin. It also explains the presence of some unusual metabolites in the urine, which could have originated from the gut microbial metabolism during the enterohepatic circulation. It has been suggested that failure of oral contraception may be due to a disruption of enterohepatic circulation following antibiotic therapy which removes the bacteria from the lower intestine that hydrolyze estrogen conjugates.

Pulmonary Excretion

The lungs are the main portal of excretion for gaseous and volatile agents such as alcohol and inhaled general anesthetic agents. The majority of these agents are excreted unchanged in the exhaled air and this allows for the anesthetic gas to be recycled in an anesthetic low flow system.

Other Routes of Excretion

Small amounts of drugs and metabolites may be excreted in sweat, saliva, tears and milk. Excretion in the saliva may impart some unique taste, such as metallic taste with metronidazole. A metabolite of rifampicin imparts an orange tinge to the sweat, tears, saliva and urine, and may sometimes be mistaken as blood in these body fluids. Milk has a slightly acidic pH, and basic drugs such as diazepam, chloramphenicol, morphine and tetracycline, are more likely to be excreted in milk, albeit in small amount. Since neonates have immature eliminating mechanisms, suckling infants may therefore be exposed to potentially dangerous levels of drugs.

Drug Clearance and Plasma Half Life

The two pharmacokinetic parameters that are related to drug elimination are clearance (CL) and plasma half-life $(t_{1/2})$. Plasma clearance of a drug is the volume of plasma cleared of the drug (by all elimination processes) per unit time. It is the sum of all the clearances $(CL_{plasma} = CL_{hepatic} + CL_{renal} + CL_{others})$, and can be calculated by the ratio of the rate of elimination to the plasma concentration.

$$CL = \frac{Rate\,of\,elimination}{Plasma\,concentration}$$

Clearance is the most important factor in determining the plasma concentration of a drug and should be taken into consideration when long-term treatment with the drug is required. Thus, when a drug is given on a regular basis,

$$\text{Dosing rate} = CL \cdot C_{p,ss}$$

where $C_{p,ss}$ is the plasma concentration at steady-state.

The general equation describing hepatic clearance is $CL_h = QE$, where CL_h, Q and E represent total hepatic drug clearance, total hepatic blood flow and the hepatic extraction ratio, respectively.

1. *High clearance, flow-dependent elimination*: For some drugs (e.g., lidocaine, morphine, nitroglycerin, propranolol, and verapamil), the extraction ratio is high (E >0.7), and the drug is removed by the liver almost as rapidly as the organ is perfused by the blood containing the drug. Consequently, the hepatic clearance of these drugs approaches hepatic blood flow and the rate of their elimination is sensitive to changes in hepatic blood flow (i.e., flow-dependent elimination). Many ingested drugs that demonstrate significant first-pass effect are drugs with high hepatic extraction ratio.

2. *Low clearance, capacity-limited elimination*: For drugs with low extraction ratio (E <0.3), the hepatic clearance is less affected by liver blood flow (i.e., flow-independent elimination). Instead the elimination of these drugs is affected more by the intrinsic activities of the metabolizing enzymes (especially the CYP 450 enzymes). Hence, drugs with low hepatic clearance display capacity-limited elimination, which is sensitive to changes in intrinsic activity (e.g., enzyme induction). Examples of low clearance drugs include phenytoin, procainamide, and theophylline.

Renal clearance is measured by the equation, $CL_r = Q_u C_u / C_P$, where CL_r, Q_u, C_u, and C_P represent total renal drug clearance, urinary flow rate, urine concentration, and plasma concentration, respectively.

Plasma half-life ($t_{1/2}$) is the time taken for the plasma drug concentration to fall to half of its original concentration. It takes 4–5 half-lives to completely eliminate a drug from the body, and it also takes the same amount of time to reach its steady state when a drug is given repeatedly. It is important to realize that the plasma half-life of a drug may change as a result of changes to either its clearance or volume of distribution according to the formula below.

$$t_{1/2} = \frac{0.693 \, V_d}{CL}$$

An increase in plasma clearance of a drug (e.g., as a result of enzyme induction) will result in a decrease of the plasma half-life of the drug, whereas an increase in the volume of distribution (e.g., pregnancy, edema) will result in an increase in the plasma half-life.

Key Concepts

- Elimination is the process that permanently removes substances from the body and this may occur physically by excretion or chemically by metabolism (biostransformation).
- Most drugs are metabolized before being excreted from the body, but polar drugs may be excreted unchanged.
- Drug metabolites are frequently more polar, more readily excreted and often biologically inactive compared to their parent compounds.
- Concurrent administration of two drugs may result in one drug affecting the rate of metabolism of the other drug.
- Renal excretion is the most important route for drug excretion.
- For weak acids and weak bases, the renal tubular pH can significantly affect the fraction of these drugs excreted unchanged in the urine.
- Enterohepatic recycling of drugs generally prolongs the duration of drug action.
- Plasma clearance of a drug is the volume of plasma cleared of the drug (by all elimination processes) per unit time. It is the most important pharmacokinetic parameter.
- Drugs with high clearance (e.g., lidocaine, propranolol, morphine) display flow-dependent elimination, whereas drugs with low clearance (e.g., phenytoin, ethanol) exhibit capacity-limited elimination.

Summary

Drug elimination may occur physically by excretion or chemically by metabolism. It is responsible for the termination of most drug actions. Drug metabolism accounts for many of the inter-individual variations seen in therapeutic drug responses, and liver is the main organ involved. Both internal and external factors may affect the rate or extent of drug metabolism. The metabolism of drugs usually results in water soluble metabolites which are readily excreted, and this generally occurs in the kidneys. The processes involved in the renal excretion of drugs are glomerular filtration, active tubular secretion and passive tubular reabsorption. While the former two processes facilitate drug excretion and are not affected by urine pH, the latter one decreases drug excretion and is pH-dependent. The two pharmacokinetic parameters that are related to drug elimination are drug clearance (CL) and plasma half-life ($t_{1/2}$). Drugs with high clearance display flow-dependent elimination, whereas drugs with low clearance exhibit capacity-limited elimination.

Further Reading

1. Begg EJ. Instant clinical pharmacology. Malden: Blackwell; 2003.
2. Boucher BA, Wood GC, Swanson JM. Pharmacokinetic changes in critical illness. Crit Care Clin. 2006;22:255–71.
3. Correia MA. Drug biotransformation. In: Katzung BG, editor. Basic and clinical pharmacology. New York: McGraw-Hill; 2012. p. 53–68.
4. Holford NHG. Pharmacokinetics and pharmacodynamics: rational dosing and the time course of drug action. In: Katzung BG, editor. Basic and clinical pharmacology. New York: McGraw-Hill; 2012. p. 37–52.
5. Hughes SG. Prescribing for the elderly patient: why do we need to exercise caution? Br J Clin Pharmacol. 1998;46:531–3.
6. Lynch T, Price A. The effect of cytochrome P450 metabolism on drug response, interactions, and adverse effects. Am Fam Physician. 2007;76:391–6.
7. Shargel L, Yu ABC. Applied biopharmaceutics and pharmacokinetics. 4th ed. New York: McGraw-Hill; 1999.

Chapter 6
Steady-State Principles

Yoo Kuen Chan and Debra Si Mui Sim

Abstract Drugs must reach adequate concentrations at the site of action to achieve therapeutic effect. This is achievable with a single dose if the desired duration is short but most times the effective concentration should be of adequate duration. In order to maintain the required concentration for adequate duration, we try to achieve steady state at the tissue level whereby the rate of infusion is equal to the rate of elimination. With constant infusion, this is usually achievable within four to five half-lives of the drug. There are several strategies to accelerate the onset of steady state both in the plasma as well as in the tissue. This includes using a loading dose or intermittent loading doses. Once the steady state is achieved, the maintenance dose must still be continued. When the maintenance dose is stopped, the concentration in the blood and tissue starts to fall and may decrease below therapeutic levels. In the critically ill patients, tissue perfusion and renal function may affect the achievement and maintenance of steady state at the target site.

Keywords Steady state • Loading dose • Maintenance dose • Elimination rate • Half-lives

Introduction

Drugs are administered with the aim of achieving adequate concentrations at the site of action for a specific response to occur. This may be achieved with a single dose, if the desired duration of action is relatively short, and the amount of drug given is sufficient to achieve the effective concentration for that duration. However, in acute care it is often desirable to ensure that the drug remains at an adequate concentration at the target site for a prolonged duration, as long as is needed for the response to occur.

Y.K. Chan, M.B.B.S., FFARCS (Ireland) (✉)
Department of Anesthesiology, Faculty of Medicine, University of Malaya, 50603 Kuala Lumpur, Malaysia
e-mail: chanyk@um.edu.my

D.S.M. Sim, B.Sc., Ph.D.
Department of Pharmacology, Faculty of Medicine, University of Malaya, 50603 Kuala Lumpur, Malaysia

© Springer International Publishing Switzerland 2015
Y.K. Chan et al. (eds.), *Pharmacological Basis of Acute Care*,
DOI 10.1007/978-3-319-10386-0_6

In order to reach this condition known as steady state, continued infusion or continued intermittent administration is required, and this administration must keep abreast with the elimination of the drug. Interestingly, many physiological processes in our body are in steady state, with the production or uptake balancing the elimination process.

How to Achieve Steady State

Steady state is achieved when the [plasma concentration]/[concentration at the site of action] is kept constant, with the rate of drug entering the body being equal to the rate of drug eliminated from the body. At the start of an intravenous infusion, the rate at which the drug enters the circulation exceeds that at which it is eliminated from the circulation. So the drug accumulates in the body. With time, the amount of drug in the circulation has accumulated to such an extent that its elimination occurs at the same rate as its entry into the circulation. That is when steady state occurs and the plasma concentration attained at this moment is known as the steady-state plasma concentration (Cp_{ss}).

For drugs administered intravenously, a constant infusion will bring it gradually to the steady state after 4–5 elimination half-lives of the drug (Fig. 6.1A). By this time, the plasma drug concentration would have reached approximately 94–98 % of its final steady-state concentration. The main determinant of the maintenance dose (A/τ) or infusion rate (R_0) to achieve the desired Cp_{ss} is the clearance (CL) of the drug, as shown in the equation below:

$$R_0 \text{ or } A/\tau = Cp_{ss}CL$$

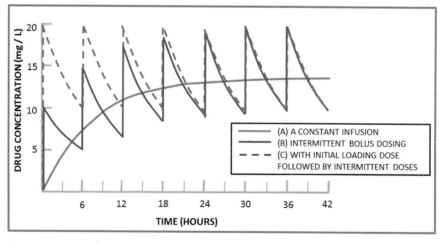

Fig. 6.1 Achieving steady state with the use of: (*A*) a constant infusion (*B*) intermittent bolus dosing (*C*) an initial loading dose followed by intermittent doses

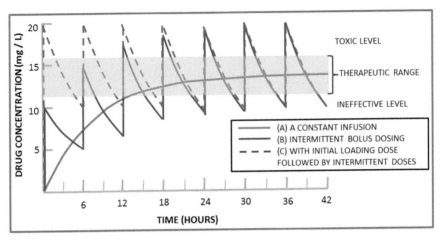

Fig. 6.2 Administering intermittent bolus doses of a drug with narrow therapeutic range expose the patient to toxic and ineffective levels during the peak and trough periods respectively

A single bolus dose (loading dose) administered before the constant infusion or intermittent bolus doses (Fig. 6.1C) can accelerate the achievement of steady state. Here, the main factor determining the loading dose (D_L) is the volume of distribution (V_d), where

$$D_L = Cp_{ss}V_d$$

If a constant infusion had been used to achieve the steady state, the variability between the maximum and the minimum plasma concentrations will be minimal. If the steady state had been achieved with bolus doses administered intermittently (Fig. 6.1B), one will see a peak and a trough concentration. For drugs with narrow therapeutic windows, the peak concentration can be in the toxic range and the trough in the ineffective range (Fig. 6.2). Thus, one may encounter a large fluctuation of responses between toxic and ineffective levels in every dosing cycle.

Increasing the frequency of the intermittent administration while reducing the dose can help to reduce the fluctuation and cause it to resemble the practice of an infusion but this is likely to be accompanied with poorer patient compliance in taking the drug.

What Happens When the Infusion Dose or the Intake Is Changed After Steady State Is Achieved

When the dose is changed after achieving steady state, a new steady-state concentration will be achieved after another 4–5 half-lives of the drug. If one wants to achieve the new steady state faster, another loading dose will bring it to the state earlier.

Tissue Concentration Versus the Plasma Concentration at Steady State

The tissue concentration is usually proportional to the plasma concentration (Fig. 6.3) at any one time. The uptake of drugs by different tissues is however different. Some drugs are polar molecules (e.g., gentamicin) and are therefore distributed mainly into extracellular water. Others are fat soluble (e.g., thiopental, diazepam) and distribute more into the fat depots in the body. So at plasma steady state, the various tissues may have reached steady state too, otherwise the plasma concentrations will not continue to be steady. These various tissues will however reach different steady-state concentrations (Fig. 6.4): some higher than that of plasma, others of the same concentration as that of plasma, and yet others at a lower concentration compared to that of plasma.

Maintenance of Steady State

Even after steady state is achieved with an infusion, infusion preceded by loading dose or infusion with intermittent loading doses, it must be maintained with a constant infusion dose to match the dose lost through the elimination process (i.e., input = output).

Steady state is no longer maintained at any site once the infusion or intermittent administrations of drug stops. When this occurs, the drug level is only affected by the process of elimination and the concentrations of the drug at the various tissue sites decline (Fig. 6.4).

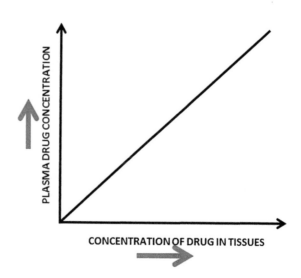

Fig. 6.3 There is a correlation between the plasma concentration and the concentration in the tissues

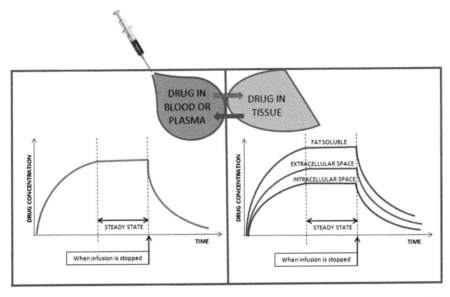

Fig. 6.4 At steady state the drug concentrations in the plasma and the tissues remain steady. The concentrations at the various tissues are different depending on the affinity of the tissue for the drug

Steady State in the Acutely Ill Patient

Many physiological processes are altered in the acutely ill patient. The variability in pharmacokinetics and pharmacodynamics is definitely greater as many of the processes involved in absorption, distribution, biotransformation and elimination would be affected.

For example, in shock states, the cardiac output is redistributed to maintain a greater proportion of blood flow to the vital organs (like the central nervous system and the heart, and to a lesser extent the gastrointestinal tract, the kidneys and liver). This means that the concentrations of drugs during steady state may be different at different tissues in these patients. In fact, the tissue concentrations of the drug at steady state may be excessively low and this may account for failure of therapy in acutely ill patients (see Chap. 25).

In those with organ failure, especially renal impairment, the blood concentrations and the tissue concentrations of the drugs are likely to be at a higher steady-state level if the conventional maintenance dose is administered. It is important in these patients to reduce the dose or to reduce the frequency at which the dose is administered, to achieve drug concentrations within the therapeutic range found in physiologically normal individuals.

Why Steady State Should Be Maintained

The action of a drug is dependent on attaining an adequate concentration of the drug at the site of the tissue for which a response is intended. This site should not only have adequate tissue concentrations of the drug but also have the tissue concentrations for a sufficient duration of time for the successful completion of the therapy.

Steady state should not only be maintained for adequate duration but should be maintained within the therapeutic range of the drug so that the benefit is maximum and the risk minimum.

Key Concepts

- The tissue concentration of a drug has to be adequate for a desired response and it has to be adequate for a specific duration.
- With continuous infusion, the steady state is usually achieved with 4–5 half-lives of the drug.
- During steady state, the infusion rate equals the elimination rate of the drug.
- Steady state needs to be maintained with continuous infusion of the drug.
- When the infusion of a drug is stopped after steady state is achieved, the steady state no longer is maintained.

Summary

Drugs must reach adequate concentrations at the site of action to achieve therapeutic effect. This is achievable with a single dose if the desired duration is short but most times the effective concentration should be of adequate duration. A constant drug concentration called the steady-state concentration is aimed for, and this is maintained by continuous infusion of the drug, at a rate which is equal to the rate of elimination of the drug. With continuous infusion, the steady state is usually achieved in four to five half lives of the drug. There are several strategies to accelerate the onset of steady state both in the plasma as well as in the tissue. This includes using a loading dose or intermittent loading doses. Once the steady state is arrived at, the maintenance dose must still be infused. When the maintenance dose is stopped, the concentrations in the blood and tissues start to decrease and may decline below therapeutic levels. In critically ill patients, tissue perfusion and renal function may affect the achievement and maintenance of steady state at the target tissue.

Further Reading

1. Boucher BA, Wood GC, Swanson JM. Pharmacokinetic changes in critical illness. Crit Care Clin. 2006;22:255–71.
2. Joukhadar C, Frossard M, Mayer BX, et al. Impaired target site penetration of beta-lactams may account for therapeutic failure in patients with septic shock. Crit Care Med. 2001;29(2):385–91.
3. Roberts F, Freshwater-Turner D. Pharmacokinetics and anaesthesia. Br J Anaesth. 2007;7(1):25–9.
4. Rondelli I, Acerbi D, Ventura P. Steady-state pharmacokinetics of ipriflavone and its metabolites in patients with renal failure. Int J Clin Pharmacol Res. 1991;11(4):183–92.

Chapter 7
Dose Response Relationship

Choo Hock Tan

Abstract Drugs act either by receptor or non-receptor-mediated mechanism. A receptor is usually a macromolecule of a cell with which an endogenous substance or a drug interacts (through specific recognition of binding domain) and elicits its effect (through transduction of signal into response). The intensity of response generally increases with plasma drug concentrations (reflected by doses administered), yielding a sigmoidal curve when the response is plotted against logarithmic values of drug concentrations or doses. The linear slope indicates the range of doses in direct proportion to the intensity of response, while the maximum indicates that receptors are fully occupied by the drug, and doses given close to and more than this response point can potentially cause overdose toxicity. A drug that is able to elicit a maximal response of a receptor is called a full agonist, while the response elicited by a partial agonist is submaximal. In the presence of a partial agonist, the effect of a full agonist can be reduced and the condition may precipitate a withdrawal syndrome for drugs like narcotic opiates. A drug that blocks the action of an agonist is called an antagonist; increasing the agonist concentration may overcome the blockade caused by a competitive antagonist (i.e. regaining the maximal response), but not that caused by a non-competitive antagonist (the response achieved is always below the maximum). Understanding of the relationship between drug dose and response as well as the effect of agonist and antagonist is important for dosing optimization. Dose-dependent adverse effect can be avoided with finely tailored drug dosages for patients, especially those with impaired organ function. Therapeutic index serves as an indicator to estimate the safety margin of a drug over a range of dose.

Keywords Drug-receptor interaction • Log dose-response curve • Full agonist • Partial agonist • Competitive antagonist • Non-competitive antagonist • Median effective dose (ED_{50}) • Median toxic dose (TD_{50}) • Therapeutic index • Therapeutic drug monitoring

C.H. Tan, M.B.B.S., Ph.D. (✉)
Department of Pharmacology, Faculty of Medicine, University of Malaya,
50603 Kuala Lumpur, Malaysia
e-mail: tanch@um.edu.my

© Springer International Publishing Switzerland 2015
Y.K. Chan et al. (eds.), *Pharmacological Basis of Acute Care*,
DOI 10.1007/978-3-319-10386-0_7

Introduction

When a drug is administered, the desired response of the drug is what most providers will aim for whilst trying to avoid the undesirable or toxic responses. For most drugs, the toxic response comes with a higher dose of the drug. Knowing the relationship between the dose and response will help a provider refine his dosing for the patient so that he consistently keeps the patient in the desirable response range.

Whilst there is no standard patient, most patients when given the usual recommended dose will respond appropriately, almost similarly to trial patients during the drug testing stage. Dose response relationship studied in populations of patients especially to look either at desirable response or undesirable response is an all or none phenomenon but offers useful information about appropriate dose to administer.

Mechanism of Drug Action

A drug is defined as a chemical when applied to a physiological system; it affects the function of the system in a specific way. Most drugs act by associating with specific 'target' macromolecules in the body to produce their effects – either a desired one (therapeutic), or an adverse one (toxic). Generally, drug receptors are classified into four major types: receptor, ion channel, enzymes, transport protein/carrier molecule. The majority of drugs act on receptors, and their actions are mediated through receptor-effector linkages, differentiated based on several distinct characteristics (Table 7.1). The ligand-gated ion channel is known to produce the fastest response (within milliseconds).

Table 7.1 The four main types of proteins as drug targets and their characteristics

	Ligand-gated ion channel	G-protein-coupled receptors	Receptor kinase	Nuclear receptors
Location	Membrane	Membrane	Membrane	Intracellular
Coupling	Direct	G-protein	Direct	Through DNA
Effector	Ion channel (resulting in membrane hyperpolarization or depolarization)	Enzyme (adenylate cyclase, phospholipase C) or channel	Protein kinases	Gene transcription
Time scale to response	Milliseconds	Seconds	Hours	Hours
Example	Nicotinic cholinoceptor	Muscarinic cholinoceptor	Growth hormone, insulin, cytokines	Steroid receptor
	GABA$_A$ receptor	Adrenergic receptor		

Note: G-protein-coupled receptors are also called metabotropic receptors

The basic concept of drug-receptor (receptor is used loosely to mean the binding target) interaction can be described by the lock-and-key model, where the structure of drug molecule influences the binding to a receptor. The affinity (the tendency of binding) of a drug for a receptor is determined by the fitting and the number of bonds formed between them. In general, the more fitted the "key and lock" feature and the number of bonds formed, the stronger the attractive forces and the binding between them. Drug-receptor binding as such is needed for drug action that produces a stimulatory or inhibitory effect as the drug response.

A few drugs act by physicochemical mechanism without the involvement of receptor. These include drugs like desferrioxamine (a chelating agent) in heavy metal poisoning, aluminium hydroxide (an antacid) to neutralize gastric acid, and mannitol (an osmotic diuretic) in treating cerebral edema.

Relationship Between Drug Response and Drug Concentration

The effect of a drug is dependent on the concentration of the drug at the target site (receptor), where drug-receptor binding is in equilibrium. The drug concentrations, in turn, depend on pharmacokinetic and dosing factors. Generally, the intensity of response increases with drug concentrations. The response can be plotted against drug concentration, yielding a hyperbolic curve that illustrates a sharp increment in effect that diminishes toward a plateau (maximal effect) while drug concentration increases (Fig. 7.1). Using this plot, unfortunately, we would not be able to visualize and appreciate the range of doses over which a linear relationship between dose and response is generated. When doses (x-axis) are converted from arithmetic to logarithmic scale, the plotted log dose–response curve will change to a sigmoidal curve where the important linear slope (25–75 % of the maximum) becomes visible for further characterization (Fig. 7.1). Therapeutic concentrations usually range along the linear portion of the curve; doses lower than the range (near threshold dose) can be sub-therapeutic, while toxic effects are likely to occur with higher doses (near the maximal effect).

Drug–Receptor Interaction: Agonism and Antagonism

Agonist

This refers to a drug that binds to a receptor eliciting its response. If the response elicited is maximal, the drug is said to be a full agonist with intrinsic efficacy of 1. A partial agonist is one that elicits submaximal response and possesses intrinsic efficacy between 0-1. An example of a full agonist for μ-receptor is the narcotic morphine, which antinociceptive and euphoric effect (referred to as maximum) can only be achieved at a lesser degree by buprenorphine, a partial agonist. The interaction of the two opioids is described in the clinical correlation.

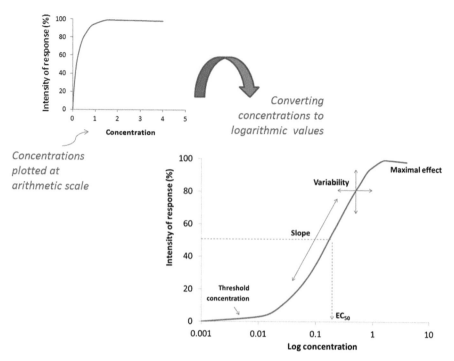

Fig. 7.1 This is a representative log concentration-effect curve, illustrating its four characteristic variables. The effect is measured as a function of increasing drug concentration surrounding the target receptor inferred from plasma drug concentration. Note that the data plotted in arithmetic scale produces a hyperbolic curve instead. $EC_{50} = $ *effective dose at half maximal response, as an indicator of potency especially useful when comparing different agonists for the same receptor.* Similar relationship can be plotted as the function of the dose of drug administered, and referred to as dose-response curve, assuming that the doses administered and plasma drug concentrations remain in equilibrium

---

Clinical correlation – Effect of a partial agonist on a full agonist

Buprenorphine dose-response curve is usually submaximal compared to morphine when used as an antinociceptive agent in acute severe pain. It is an opioid drug with similar affinities to both μ- and κ- receptors, possesses intermediate efficacy at the μ-receptor (partial agonism), but antagonizes the activity of κ-receptors. The clinical significance of buprenorphine as a partial opioid agonist is explained essentially by its submaximal effect on μ-receptor: it can be used as a potent analgesic with a lower risk of abuse, addiction, and side effects compared to a full opioid agonist like morphine; in addition at low doses it is also a treatment for opioid addiction without causing significant withdrawal symptoms. However, being a partial agonist, it can (especially at high doses) provoke a withdrawal in those physically dependent, who has just taken their usual "high doses" of heroin or morphine – a classical illustration of a normally useful partial agonist that has turned offensive by antagonizing the effect of a full agonist, bringing on the withdrawal syndrome.

A less typical agonist type called 'inverse agonist' is one that binds to a receptor but dose-dependently suppresses the constitutive activity (basal level of activity without any ligand present) of the receptor. An example is metoprolol, though widely known as a β-adrenergic antagonist, is in fact also an inverse agonist at the adrenergic receptor, reducing the sympathetic activity of the heart for the optimization of heart failure management.

Antagonist

An antagonist is a drug that binds a receptor and prevents its activation by agonist (resulting in a flat response). The effect of drug antagonism can be achieved via:

(i) **Competitive antagonism**: antagonist and agonist compete for receptor binding site. Competitive agonist lacks intrinsic efficacy (0) but retains affinity for receptor's binding site. This produces a characteristic concentration-dependent parallel shift to the right of the dose–response curve without altering the maximal response. Clinically, the anticipated response at a given dose of agonist drug will not occur. However, the response can be regained once the agonist concentration is increased (i.e. reversible block) [Fig. 7.2]. A competitive agonist with beneficial use in acute care of opiate overdose is exemplified by naloxone in the following clinical correlation.

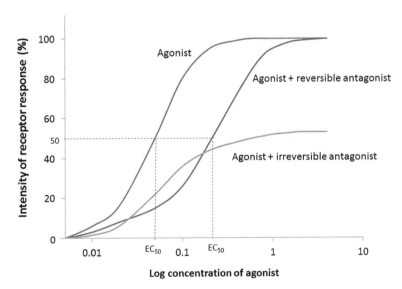

Fig. 7.2 Comparison of the effects of reversible (competitive) and irreversible (non-competitive) antagonists on the response induced by an agonist

---

Clinical Correlation – Effect of Naloxone in Drug Addicts

Naloxone binds with high affinity to μ-opioid receptors in the central nervous system and rapidly blocks the action of opioid narcotics on the receptors, hence reversing the respiratory depression. Administering this drug to drug addicts however can precipitate a withdrawal syndrome rapidly, more potent than that caused by the partial agonist buprenorphine. The reduction in the μ-receptor-mediated effect sustained by morphine essentially causes the withdrawal syndrome in these patients.

(ii) **Non-competitive antagonism**: In the presence of a non-competitive antagonist, an agonist fails to elicit maximal response despite increasing concentrations. Two mechanisms have been described: (a) pseudo-irreversible block: the affinity of antagonist is so strong that it binds very tightly via covalent bond to the receptor. Its slow dissociation from the target site prevents the agonist from binding to the receptor; (b) true irreversible block: the antagonist binds to a receptor site (allosteric site) different from that targeted by agonist, resulting in a conformational change in receptor that renders it unfavorable for agonist binding. On the concentration-time curve, the agonistic response is achieved below maximum even when its concentration increases (Fig. 7.2).

Relationship Between Drug Response and Drug Doses in the Population

Most adverse effects can be viewed as an extension of a drug's pharmacological action, secondary to accumulation or overdose of a drug in the body exceeding the maximal tolerable concentration. Dose-dependent toxic drug effects are usually studied in a population during drug trials as a quantal drug response (the drug effect either occurs, or it does not). Quantal dose-response curves are useful to estimate doses to which most of the population responds – both therapeutic and toxic responses to a drug [Fig. 7.3]. In the instance of the opioid drugs, the dose range between the median analgesic dose and median respiratory depressant dose is a 'safety margin' which indicates the likelihood that the toxic effect will occur – generally, the wider the margin, the safer a drug is. Most forgiving among the opioid analgesics is probably remifentanil, which offers a therapeutic index of 33,000:1, while morphine has an index of 70:1 (still considered safe). Drugs such as digoxin and phenytoin have very narrow therapeutic indices of approximately 2:1, mandating close monitoring of therapeutic and toxic effects, either by clinical assessment or investigation (plasma drug monitoring).

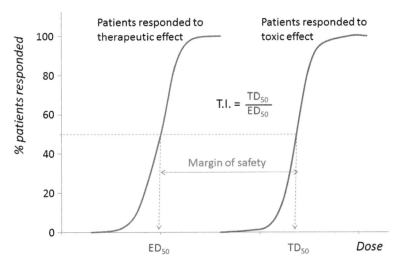

Fig. 7.3 A log dose-response curve depicting the percentage of patients responding to the therapeutic effect and the toxic effect of a drug. Drug toxicity develops at higher doses, indicating it as an extension of pharmacological action of the drug on the body. ED_{50} = median effective dose; TD_{50} = median toxic dose; TI = therapeutic index

---

Clinical Correlation – Therapeutic Drug Monitoring

One of the criteria in necessitating therapeutic drug monitoring of a drug is the safety profile of the drug. The safety profile can be determined by the therapeutic index (TI) of the drug; essentially, the lower the TI, the more toxic a drug can be. Drugs that require therapeutic monitoring to balance its therapeutic effect with minimal toxicity include theophylline (cardiac and neurological toxicity), gentamycin and vancomycin (renal and ototoxicity). Therapeutic drug monitoring depends greatly on the pharmacokinetic profile and hence sampling time is of utmost importance to provide accurate information for valid interpretation. It can also be applied in acute acetaminophen poisoning, where the plasma level of acetaminophen interpreted with the time course of ingestion indicates the need to initiate antidote (acetylcysteine) treatment.

Key Concepts

- Drug response is usually elicited by drug-receptor interaction.
- Administration of a partial agonist or antagonist essentially reduces the maximal response sustained by a full agonist, and can cause withdrawal symptoms.
- Doses should be tailored and fine-tuned to keep within the therapeutic range especially for drugs with a narrow therapeutic window.

- Monitoring and dose adjustment is particularly important in the critically ill patient with impaired organ functions because the pharmacokinetics and pharmacodynamics of drugs are altered even if the drug concentration is in the normal therapeutic ranges.

Summary

The majority of drugs act through binding of receptors. The presence of agonists or antagonists can alter the dose (concentration)-response relationship. Adverse or toxic drug effects often occur with high doses of a drug, manifested as an extension of the pharmacological action of the drug. The ratio between median effective dose (ED_{50}) and median toxic dose (TD_{50}) indicates the margin of safety of a drug, and is useful as an indicator of the relative safety profile of different drugs. Close monitoring of response to guide adjustment of dosages is especially important in those who are acutely ill as their pharmacokinetic and pharmacodynamic profiles are altered.

Further Reading

1. Brunton L, Blumenthal D, Buxton I, Parker K, editors. Goodman & Gilman manual of pharmacology and therapeutics. 11th ed. New York: McGraw-Hill; 2008.
2. Harvey RA, Clark MA, Finkel R, Rey JA, Whalen K, editors. Lippincott illustrated reviews pharmacology. 5th ed. Philadelphia: Wolters Kluwer; 2012.
3. Holford NHG. Pharmacokinetics and pharmacodynamics: rational dosing and the time course of drug action. In: Katzung BG, editor. Basic and clinical pharmacology. New York: McGraw Hill; 2012. p. 37–52.
4. Rang HP, Dale MM, Ritter JM, Flower RJ, Henderson G. Rang and Dale's pharmacology. 7th ed. Edinburgh: Churchill Livingstone; 2012.
5. von Zastrow M. Drug receptors and pharmacodynamics. In: Katzung BG, editor. Basic and clinical pharmacology. New York: McGraw-Hill; 2012. p. 15–36.

Chapter 8
Pharmaceutical Aspects of Drugs

Pauline Siew Mei Lai

Abstract Drugs can be obtained from many sources such as plants, animals, minerals or microorganisms. However, most drugs are synthetically produced in the laboratory, as they can be produced on a larger scale, in a more cost effective manner and be of higher quality. It is important to know the sources of drugs as patients can develop hypersensitivity reactions to certain drug source. Some patients (based on their religious beliefs) also prefer not to use drugs obtained from bovine or porcine sources. Drug development and trials in animals and humans is a long and costly process. This process is essential to determine that the new drug is safe for human consumption. Drugs should be packaged in a manner that protects the active ingredient from deterioration due to external factors. Drugs should also be labeled with sufficient information to enable the determination of the exact content of the active ingredient, its storage conditions and manufacturing details.

Keywords Drug sources • Drug development • Clinical trial • Packaging • Labelling

Introduction

Drugs can be obtained from many sources, such as plants, animals, and minerals. Today, most drugs are synthetically manufactured in laboratories, or produced by microorganisms. Most drugs have undergone several years in the developmental stages where intensive research, drug trials and safety testing on them had been done on the generic form. After they have been cleared for human consumption through all these testing and post testing surveillance, the company that has invested in the research would commission for its production and registration with national and international drug registries. Drugs as it is available to the patients are also packaged in such a way that the original drug is protected from the many agents that can contribute to its deterioration.

P.S.M. Lai, B.Pharm., Ph.D. (✉)
Department of Primary Care Medicine, Faculty of Medicine, University of Malaya,
50603 Kuala Lumpur, Malaysia
e-mail: plai@ummce.edu.my

© Springer International Publishing Switzerland 2015
Y.K. Chan et al. (eds.), *Pharmacological Basis of Acute Care*,
DOI 10.1007/978-3-319-10386-0_8

Sources of Drugs

Plants as Sources of Drugs

Plants have been used as drugs as far back as 2700 B.C. Drugs can be sourced from most parts of a plant. Atropine and caffeine are obtained from leaves; castor oil and strychnine are extracted from seeds; hyoscine and quinine are extracted from barks; morphine and vincristine are extracted from flowers; ipecacuanha and reserpine are extracted from roots; tubocurarine is extracted from stems; and anthracene and physostigmine are extracted from fruits.

Today, plants are no longer used in the raw form. Instead, their active ingredients are extracted and identified for its pharmacodynamics and pharmacokinetic properties. This is performed to ensure that a precise and constant dosage is available for therapeutic use. Scientists will also look into the possibility of chemical synthesis for sustainability. Active ingredients obtained from plants can be grouped according to their physiochemical properties.

Alkaloids

An alkaloid is defined as a "basic nitrogenous compound of plant origin which produces salt when combined with acid, and is physiologically active in the plant and animal". Alkaloids are a white crystalline substance, bitter and insoluble in water. Its salt preparation however, is highly soluble in water. Alkaloids can be classified according to their plant source. For example: belladonna (atropine), cinchona (quinine), cocaine, ergot (ergotamine), opium (morphine), rauwolfia (reserpine), vinca (vincristine), and xanthine (caffeine). Names of alkaloids end in "...ine".

Glycosides

Glycosides are non-nitrogenous, colorless, crystalline solids. They can be split into sugar and non-sugar parts with the addition of an acid or enzyme. They do not form salts. Some glycosides are poisonous. The non-sugar part of glycosides (called aglycone) is responsible for its pharmacological activity. Chemically, they are similar to bile acid, sterol and steroid hormones. When sugar is combined with aglycone, the lipid/water partition coefficient, potency, and pharmacokinetic properties are modified. Glycosides can be classified as glucoside, galactoside, fructoside according to the presence of these sugars. Glycosides are found in the bark, seed and leaf of a plant. An example is Digoxin, which can be isolated from the leaves of the purple foxgloves, Digitalis purpurea.

Oils

Fixed oils are a mixture of glycerol esters (such as palmitic, stearic, and oleic acid). They are not volatile, are lighter than water, are insoluble in water, but are soluble in chloroform and ether. They are not dissipated by heat. Olive oil and castor oil are examples of fixed oils. Metabolites of castor oil irritate the mucosa of the gastrointestinal tract, which promotes peristalsis and gut evacuation. Olive oil is edible. It can also be used as an emollient.

Volatile oils are the odorous part of a plant. They are sometimes called essential oils, as they represent the essence of a plant. When fresh, volatile oils are colorless. When exposed to the environment, they turn dark. They can evaporate. They can also be oxidized and resinified. Hence, volatile oils should be stored in a cool, dry place, tightly closed in an amber glass container. Peppermint oil and spearmint oil are examples of volatile oils. Menthol is the active ingredient of peppermint oil. These oils are used as solvents and flavors in the compounding of preparations.

Gums and Mucilage

Gum is a secretory hydrocarbon product which originates from plants. When hydrolysed, it produces sugar. Gum dissolves in water easily, whilst mucilage forms a slimy mass. Agar is an example of gum. When swallowed, gum absorbs water to form bulk, thereby exerting its laxative effect. Tragacanth is an example of mucilage. It is most commonly used as a suspending agent for insoluble powder in mixtures, an emulsifying agent for oil and resin, or as an adhesive.

Carbohydrate and Related Compounds

Sucrose is used as a demulcent and nutrient. Concentration of sucrose of more than 65 % is bacteriostatic and acts as a preservative. Dextrose is a nutrient, which may be given orally or intravenously. Fructose is used in food, particularly for patients with diabetes. Alcohol at 70 % concentration is used as an antiseptic.

Animals as Sources of Drugs

Insulin was initially extracted from the pancreas of cows (bovine) or pigs (porcine). Heparin is extracted from porcine intestinal mucosa or bovine lung. Human menopausal gonadotropins are isolated from the urine of postmenopausal women. However, animals as sources of drugs are frequently associated with hypersensitivity reactions. Some patients may also prefer not to have drugs obtained from porcine or bovine sources.

Minerals as Sources of Drugs

Clay, kaolin and activated charcoal are used for diarrhea. Iodine is used for the treatment of goiter. Gold is used for arthritis. Externally, sulfur is used for treating skin diseases. Most antacids contain aluminum hydroxide and magnesium trisilicate. To relieve constipation and to control eclamptic seizures, magnesium sulphate can be used.

Laboratory as Sources of Drugs

Today, most drugs are artificially synthesized as it is safer, and more effective than extracting drugs from plants or animals. Drugs produced in laboratories are of high quality and less expensive. They can be produced on a large scale within a short period of time. Examples are digoxin and insulin.

Microorganisms as Sources of Drugs

Drugs produced by microorganisms include amphotericin and chloramphenicol.

Clinical Correlation – Implications of Sources of Drugs

Some drugs (especially those sourced from animals) are frequently associated with hypersensitivity reactions. Before administering the drug, it is necessary to ascertain that the patient does not have any allergy to animal sources. Viper venom antiserum for example, is produced in horses (equine).

For religious reasons, some patients may also prefer not to have a drug from a porcine (relevant for Muslims patients) or bovine (relevant for Hindu patients) source. Alternative drug sources will need to be found for these patients.

Drug Development

Drug development is defined as the process of bringing a new drug to the market once a compound has been identified as potentially being able to impact a particular physiological/pharmacological process in the body. There are two phases to drug development.

The Pre-clinical Phase

Compounds that emerge from the process of drug discovery are called new chemical entities (NCEs). They will have some promising activity against a biological target, in a certain disease. However, little is known about the safety, toxicity, pharmacokinetics and metabolism of this compound in humans. Hence, these parameters will need to be assessed prior to human clinical trials. The dose and frequency of dosing will also need to be determined. The physicochemical properties of this new chemical entity (its chemical makeup, stability, solubility) will need to be established. Production on a large scale to different pharmaceutical formulations (known as chemistry, manufacturing and control [CMC]) will also need to be performed.

Many aspects of drug development are to satisfy regulatory requirements of drug licensing authorities to ensure safety. A number of tests need to be conducted to determine major organ toxicities of the new compound prior to its use in humans. Some tests can be performed using in vitro methods. However, many tests still require the use of experimental animals, to examine the complex interplay of metabolism and effects of exposure to toxicity.

Information obtained from the pre-clinical phase and CMC is submitted to regulatory authorities as an Investigational New Drug (IND) application. If the new drug is approved, development moves to the clinical phase assessing the efficacy and safety of the drug.

Clinical Phase

The process of drug development continues even in human clinical trials (Table 8.1). Long-term or chronic toxicities are determined, as well as its effects on fertility, reproduction and other systems, and its cancer causing effects if any. If the compound has an acceptable toxicity and safety profile, and the desired effect

Table 8.1 The various phases of drug trials through human subjects

Phase of drug trial	What is being done
Phase I trials	Conducted in healthy volunteers, to determine safety and dosing
Phase II trials	Used to get an initial reading of efficacy and further explore safety in a small numbers of sick patients
Phase III trials	Large, pivotal trials to determine the safety and efficacy in patients
Phase IV trials (optional)	Post-market surveillance studies, to assess the safety and efficacy of the drug after its release to the market

in clinical trials, it will be submitted for marketing approval. The drug then becomes ready for registration and distribution to pharmacies. However, most new compounds fail these tests. The success rate for a new chemical entity to successfully complete Phase I-III clinical trials to the point of registration is low.

Drug Packaging

Packaging is defined as the "collection of different components which surround the pharmaceutical product from the time of production until its use". The quality of the packaging of the pharmaceutical products plays an important role. It must

1. Protect against all adverse external influences that can alter the properties of the product (moisture, light, oxygen and temperature variations)
2. Protect against biological contamination
3. Protect against physical damage
4. Have the correct information and identification of the product

The packaging of materials used must be chosen so that the packaging does not have an adverse effect on the product and vice versa. Written labels on the packaging must include the following information: the international non-proprietary name (INN) of the active ingredient, the dosage form and trade name, information on storage and the batch number. Packaging and labeling allows compliance with the Code of Good Manufacturing Practice.

Clinical Correlation – Importance of Proper Storage Conditions for Drugs

Drugs may need special storage conditions to prolong their shelf life and require special packaging.

Co-amoxiclav has two active components: amoxicillin and clavulanic acid. Amoxicillin is sensitive to dehydration, whilst clavulanic acid is sensitive to moisture. An imbalance in dryness or moisture can deactivate one or the other of the active components of the drug, thereby, affecting potency of the components of the drug. The packaging of co-amoxiclav needs to be specially designed to maintain the efficacy of the two active ingredients that require contrasting protective needs. The innovator pharmaceutical company of co-amoxiclav has packed the drug in a desiccated pouch with material that helps maintain just the right amount of moisture ideal for both components during transport and storage.

The generic companies may not consider this fact in their packaging and this may explain the reduced potency of the drug from these companies.

Key Concepts

- Drugs can be obtained from many sources. However, presently, most drugs are manufactured synthetically in laboratories.
- Drug development consists of two phases: preclinical and the clinical phase.
- The preclinical phase determines the physicochemical properties of the compound, possible production to a large scale, and its toxicity profile.
- The clinical phase involves clinical trials on humans to determine its safety profile.
- Packaging should protect against all adverse external influences that can alter the properties of the product, biological contamination and physical damage.
- Packaging should also carry the correct information and identification of the product.

Summary

It is important to know the source of a drug, as animal sources of drugs are frequently associated with hypersensitivity reactions. For religious reasons, some patients may also prefer not to have a drug from a porcine or bovine source. Drug development is a slow and costly process. It takes several years before a new chemical entity passes successfully through all the stages of clinical trials and is registered with the relevant authorities for sale in a country. Good drug packaging is required to protect the drug from the environment, and to protect the environment from the drug. All packaging has to be labeled appropriately according to the Code of Good Manufacturing Practice.

Further Reading

1. Adams CP, Brantner VV. Spending on new drug development. Health Econ. 2010;19(2):130–41.
2. Liljefors T, Krogsgaard-Larsen P, Madsen U, editors. Textbook of drug design and discovery, Third Edition (Forensic Science). 3rd ed. New York: Taylor & Francis; 2002.
3. Pharmaceutical Inspection Co-operation Scheme Secretariat. Guide to good manufacturing practice for medicinal products. Geneva 2009. Available from: http://www.tga.gov.au/pdf/manuf-pics-gmp-medicines-annexes.pdf. Last accessed 13th June 2014.
4. World Health Organization. WHO Technical Report Series, No. 902, 2002, Annex 9, Guidelines on packaging for pharmaceutical Products. Geneva 2002.

Chapter 9
Dosage Forms, Drug Calculations and Prescription

Pauline Siew Mei Lai and Yoo Kuen Chan

Abstract The pharmacodynamics and pharmacokinetics of a drug play an important role in how the drug is formulated, and which dosage form is suitable for use. It is also necessary to have different dosage forms to meet the different requirements of the patient. Drug formularies are a list of medicines available in an institution or country. When drugs are prescribed from a drug formulary, the drugs are very likely to be available. A clearly written prescription allows the pharmacist to dispense the drug without error. Error from this is further reduced by electronic prescribing which incorporates an inbuilt decision support strategy. Drug doses are expressed in different units and may be confusing to providers. An understanding of what these units signify is essential to enable accurate and reliable calculation of drug doses. It is also important to have a good understanding of this to know how to dilute the drugs and to be able to administer the correct dosage from these dilutions.

Keywords Dosage forms • Dose calculations • Drug prescription • Drug formularies • Electronic prescribing

Introduction

Dosage forms refer to the form the pharmaceutical product takes when it is available for use by patients. It may exist in different forms to meet the requirements needed for the peculiarities of storage conditions, its various routes of administration, and to facilitate its arrival to the site of action in a form that is still active.

P.S.M. Lai, B.Pharm., Ph.D. (✉)
Department of Primary Care Medicine, Faculty of Medicine, University of Malaya,
50603 Kuala Lumpur, Malaysia
e-mail: plai@ummc.edu.my

Y.K. Chan, M.B.B.S., FFARCS
Department of Anesthesiology, Faculty of Medicine, University of Malaya,
50603 Kuala Lumpur, Malaysia

© Springer International Publishing Switzerland 2015   73
Y.K. Chan et al. (eds.), *Pharmacological Basis of Acute Care*,
DOI 10.1007/978-3-319-10386-0_9

It is important to ensure that the right amount of drug arrives at the tissue to provide for optimal therapeutic effect. Toxic amounts administered as a result of faulty calculations of dosages can cause harm and possibly even death. It is therefore vital that providers understand the various terms used in the pharmaceutical industry and learn how to calculate the correct dosage for their patients especially those with abnormal physiological processes.

Dosage Forms

Enteral Route

These medications go through the gastrointestinal tract, are absorbed into the bloodstream and metabolized by the liver. Whilst most preparations taken orally may be in the form of tablets, the capsule forms are designed to delay absorption in the upper gastrointestinal tract. Liquid forms are usually made to allow children who are too young to take tablets to ingest the drugs. Suppositories allow rectal administration and are particularly useful for uncooperative children and also those who are vomiting and hence would not be able to ingest the medication.

Parenteral Route

Parenteral medications are injected directly into the bloodstream, or into body tissue. They do not pass through the liver. Injectable drugs come in the form of solutions or powders. Powders need to be reconstituted with a sterile diluent before it can be injected. These include intravenous (IV), subcutaneous (SC), intramuscular (IM), intrathecal (IT), intraperitoneal (IP), intradermal (ID), and intraosseous (IO) routes. These drug forms have to be sterile to reduce the likelihood of infection.

Inhalational Route

In the inhalation route of administration, drugs are breathed in through the mouth or the nose. It acts directly on the respiratory system before entering the bloodstream. Devices used include the inhaler or the nebulizer which allows maximum amount of active drugs to be administered to the tissues of the respiratory tract or alveoli where they can produce the desired effect.

Topical

Topical drugs are applied to the skin surface or a mucous membrane. These include creams, lotions, ointments, ear drops, eye drops, transdermal patches and vaginal rings, which upon application provide maximum concentration at the sites of need directly.

Clinical Correlation – Understanding Importance of Dosage Forms

It is important to understand the pharmacodynamics and pharmacokinetics of the drug to ensure its effective use. Certain medications like slow release nifedipine should never be crushed. If nifedipine is crushed and administered, it could result in life-threatening hypotension as the bioavailability of nifedipine is increased.

Formulary of Drugs

A formulary is a list of medicines. It is a collection of formulas for the compounding and testing of medications. A reference book frequently referred to as the "pharmacy bible" is the Martindale and the British Pharmacopoeia.

Most hospital institutions will use a prescription formulary. The main function of a prescription formulary is to specify which medications can be prescribed by which specialist. The development of prescription formularies is based on evaluations of efficacy, safety, and cost-effectiveness of drugs. Depending on the individual formulary, it may also contain additional clinical information, such as indications for the use of the drug, side effects, contraindications and recommended doses.

Drug Calculations

Patients vary in their need for drugs. They also vary in the amount of drugs needed. Providers can and do make mistakes in calculating the correct amount needed by their patients because there is great variation in not only the units of measurement of doses of drugs, but also in the way concentrations of drugs are represented.

Units of Drug Doses

Some drug doses are given in grams (g), milligrams (mg) or micrograms (mcg). Others are given in litres (L) or milliliters (ml or mL) of known concentrations, or in millimoles (mMol or mmol), units or international units. It is advisable that these units be written clearly.

Table 9.1 The two most common formulae for determining body surface area from the weight and height of a patient

The Mosteller formula
BSA (m²) = { [Height (cm) x Weight (kg)] / 3,600}$^{1/2}$
The DuBois and DuBois formula
BSA (m²)= 0.20247 x Height (m)$^{0.725}$ x Weight (kg)$^{0.425}$

Calculation of Doses Against Body Weight or Surface Area

Drug doses are usually prescribed according to body weight (per kg) or occasionally according to body surface area (per m²). When body surface area is used in drug dosing, it is important to determine the surface area of the patient before making the calculation.

Mistakes can sometimes be made whilst converting the doses from microgram to milligram or even to grams. It is therefore important to know how the various units relate to each other (1 gram = 1,000 mg; 1 mg = 1,000 mcg). It is recommended not to use decimal points in drug dosages e.g. 0.1 mg of fentanyl is better written as 100 mcg of fentanyl. So when calculating the fentanyl requirement for a 75 kg man where the recommended dose is 2 mcg/kg, the total dose required is 150 mcg of the drug.

Pediatric Doses

Pediatric doses of drugs in particular need to be carefully calculated. Sometimes the drug is prescribed according to body surface area (per m²). Under these circumstances, it is important to determine the body surface area of the child from his height and weight using standard acceptable conversion formulae (see Table 9.1) and tables.

Calculation in Concentrations

Most times the concentrations are expressed as weight/volume, eg fentanyl is 50 mcg / ml. So the volume of fentanyl required in the 75 kg man mentioned earlier is 3 ml of the 50 mcg/ml fentanyl (as 3 ml will contain 150 mcg).

Drug concentrations may sometimes be expressed as percentages (%). They can mean weight/weight (w/w %), weight/volume (w/v %), volume/volume (v/v %) or part/part percentages. The denominator in each fraction could either stand for the solvent, or the solution as a whole. When presented in this way a 5 % solution means there are 5 g of the drug in 100 g of solution (water is a frequently used solvent). As the density of water is 1 g/ml, it also means 5 g of the drug in 100 ml of the solution.

Local anaesthetics are always quoted as percentages e.g. 0.75 % ropivacaine. It means there are 0.75 g in 100 g of solution or 0.75 g in 100 ml or 750 mg in 100 ml or 7.5 mg per ml.

It is important to remember that dosages of drugs must always be in weight i.e. mg or grams. If volumes are quoted as doses, the concentrations must be known. It is wrong to order a certain volume of a drug to be administered, without first specifying the concentration of the said drug. The exception to this is with respect to gases, which are delivered in litres or ml per minute (flow rate). Gases have a fixed volume at standard temperature and pressure (at sea level) and it is understood that 2 L of oxygen from the flowmeter means 2 L/min.

Calculation in Ratios

When a drug concentration is represented as a ratio, e.g. 1:10 it means 1 g of the drug is found in 10 g of the solution or as most drugs are diluted in water (with a density of 1 g per ml), 1 g in 10 ml of water or 1,000 mg in 10 ml or 100 mg per ml. Adrenaline comes in 1:1,000 in the ampoule. It means 1 g is found in 1,000 g (1,000 ml) of solution or 1,000 mg in 1,000 ml or 1 mg per ml.

Diluting Drugs

Adrenaline from the ampoule without dilution is termed the neat dose, and is referred to as a 1:1000 solution. We usually dilute it with another 9 ml of normal saline or water so that it becomes 1 mg in 10 ml or 1 mg in 10 g of water. This is equivalent to 1 g of adrenaline in 10,000 g ($10 \times 1,000$) of water and therefore becomes a 1:10,000 solution of adrenaline.

Sometimes we are required to dilute local anaesthetics for the purpose of administration into the epidural space for analgesia. In order to dilute 0.75 % ropivacaine (7.5 mg per ml) to a 0.1 % ropivacaine solution (1 mg per ml), into a 50 ml syringe, the following can be done:

0.75 % ropivacaine contains 7.5 mg per ml; so in the 50 ml syringe you would need 50 mg of ropivacaine. Draw up 7 ml of the 0.75 % ropivacine (52.5 mg of the drug) and dilute this to a total of 52.5 ml (or 50 ml is close enough) with saline to obtain approximately 50 mg in 50 ml or 0.1 % ropivacaine solution.

Clinical Correlation – Preventing Errors with Drug Dilutions

Medication errors may occur as a result of administering drugs based on faulty dilutions. This can be very serious especially when managing patients in the pediatric age group. It is a good practice to counter check with another colleague the calculations and/or the final dose required.

Writing the Prescription

A prescription is an order by a health care provider for the pharmacist to provide medication to the patient in clear terms, detailing the dosage, the route of administration, the frequency of administration and the total duration of the administration. If written, it has to be legible so that no mistakes will arise from the process.

The name of the patient, the age and the date of the prescription should be clear. It is important to use the generic name of the drug and indicate the concentration or strength. The dosage form should also be written in case the drug has more than one dosage form. The amount to be administered per intake, the frequency and total amount to be provided (or the duration of the therapy) need to be clearly specified too. Further instructions or warnings should be included e.g. to be taken before or after a meal, especially in medications that may cause gastric discomfort when taken on an empty stomach.

The signature and the credentials of the prescriber should also be legible. This is to ensure the prescription is a bona fide document. It also allows the pharmacist to counter check with the prescriber in case of errors in the document.

In many modern hospitals, electronic prescription has taken over the written prescription. The time taken for the pharmacist to receive the prescription is reduced. As there is an inbuilt decision support strategy within the prescribing system, medication errors due this process can be reduced. Dosage, frequency, form and strength of the drug are preset. This makes it easier for the prescriber to "write" the prescription.

Key Concepts

- Many different dosage forms exist to meet the various needs of the patient.
- Most of the information about drugs can be found in drug formularies, whether in-hospital, national or international.
- Drug doses are expressed in many different units and a good understanding of this allows one to calculate the necessary doses required.
- A properly written prescription allows the appropriate dispensing of drugs to the patient.

Summary

Various dosage forms are required for different clinical situations. Most information regarding drugs is available in formularies. It must be recognized that drug doses may appear in different units, so providers must be able to understand these units and learn how to calculate the actual doses required by the patient. The pharmacist through a properly written prescription is then required to dispense the drug to appropriately meet the needs of the patient.

Further Reading

1. Brayfield A. Martindale: the complete drug reference. 38th ed. London: Pharmaceutical Press; 2014.
2. British Pharmacopoeia Commission Secretariat of the Medicines and Healthcare Products Regulatory Agency: The British Pharmacopoeia London; 2014.
3. De Vries TPGM, Henning RH, Hogerzeil HV, Fresle DF. Guide to good prescribing. Geneva: World Health Organization; 1994. WHO/DAP/94.11.
4. DuBois D, Dubois EF. A formula to estimate the approximate surface area if height and weight be known. Arch Int Med. 1916;17:863–71.
5. Mosteller RD. Simplified calculation of body surface area. N Engl J Med. 1987;317(17):1098.
6. Rolfe S, Harper NJN. Ability of hospital doctors to calculate drug doses. BMJ. 1995;310:1173.
7. Schier JG, Howland MA, Hoffman RS, Nelson LS. Fatality from administration of labetalol and crushed extended-release nifedipine. Ann Pharmacother. 2003;37(10):1420–3.
8. Simpson CM, Keijzers GB, Lind JF. A survey of drug-dose calculations skills of Australian tertiary hospital doctors. Med J Aust. 2009;190:117–20.
9. Trissel LA. Handbook on injectable drugs. American Society of Health-System Pharmacists – Online via Medicines Complete. Available from http://www.pharmpress.com/product/MC_ HID/handbook-on-injectable-drugs. Accessed May 27th, 2014.

Chapter 10
Drug Interactions

Debra Si Mui Sim and Choo Hock Tan

Abstract Drug interactions, resulting from multiple drug therapies, can be attributed to (i) physicochemical factors, e.g., incompatibility of pharmaceutical preparations, (ii) pharmacokinetic factors, e.g., interference of absorption, distribution, metabolism and excretion, or due to (iii) pharmacodynamic factors, e.g., interference at the target site or the associated signaling mechanism. Drug interactions may also occur between drugs and other xenobiotics such as food constituents, environmental pollutants and herbal products. These drug interactions produce dichotomous clinical outcomes: one that is beneficial, and one that is harmful. In many acute conditions where patients have already ended up with some compromised organ functions, both pharmacokinetic and pharmacodynamic factors can exacerbate the drug interaction problem and potentially result in toxicity. Pharmaceutical incompatibility of drug injection solutions is another potential cause of adverse drug interactions which should not be overlooked. A lack of knowledge on the various drug interactions is a major contributing factor to adverse drug reactions or therapeutic failures.

Keywords Drug interactions • Pharmaceutical incompatibility • Pharmacokinetic interaction • Pharmacodynamic interaction

Introduction

Drugs are given in combination for two reasons: (1) different conditions demanding different pharmacological agents; (2) manipulation for therapeutic advantage when their beneficial effects are additive or synergistic, or because the use of submaximal doses of individual drug lead to fewer drug-specific adverse effects yet allowing the desired effect to be achieved. Combination therapy is practiced to optimize the management of many conditions, acute or chronic, such as heart failure, diabetes mellitus, malignancy and sepsis. The resultant effect(s) from multiple drugs, i.e., drug-drug interaction, can be however beneficial or harmful. Drug interactions can be

D.S.M. Sim, B.Sc., Ph.D. (✉) • C.H. Tan, M.B.B.S., Ph.D.
Department of Pharmacology, Faculty of Medicine, University of Malaya,
50603 Kuala Lumpur, Malaysia
e-mail: debrasim@ummc.edu.my

© Springer International Publishing Switzerland 2015
Y.K. Chan et al. (eds.), *Pharmacological Basis of Acute Care*,
DOI 10.1007/978-3-319-10386-0_10

attributed to (i) physicochemical factors, e.g., incompatibility of pharmaceutical preparations, (ii) pharmacokinetic factors, e.g., interference of absorption, distribution, metabolism and excretion, or due to (iii) pharmacodynamic factors, e.g., interference at the target site or the associated signaling mechanism. Drug interactions may also occur between drugs and other xenobiotics such as food constituents, environmental pollutants and herbal products. A lack of knowledge on the various drug interactions is a major contributing factor to adverse drug reactions or therapeutic failures.

Physicochemical/Pharmaceutical Interactions

This occurs outside the body (in vitro), such as in the syringe or infusion bottle, and is often the result of pharmaceutical incompatibility. For example, while penicillin and aminoglycoside antibiotics have synergistic antimicrobial effect in vivo when given together, they must not be mixed in the same syringe as this will result in a reduction of their antimicrobial activities, caused by acid–base type of reaction. A similar interaction occurs between thiopental sodium and succinylcholine hydrochloride or vecuronium hydrochloride if given in the same intravenous mixture. These solutions possess physicochemical properties that are not compatible and will form precipitates when they are mixed in vitro.

Furthermore, the problem of pharmaceutical incompatibility is frequently observed when drugs which are poorly water soluble are formulated as an injection solution in some specialized injection vehicle (e.g., diazepam in 50 % propylene glycol and 10 % ethanol; phenytoin sodium formulated with non-aqueous solubilizing agents and adjusted to a pH of 12) and these drugs may precipitate when the injection solution is diluted in aqueous solution or mixed with a solution of different pH. Consequently, diazepam injection solution may form precipitate if diluted in normal saline (pH 7.0–7.5), whereas phenytoin sodium injection solution may lose its activity when added to an infusion bag of glucose 5 % solution (pH 4.3–4.5). Other drugs with similar solubility problem include digoxin, clonazepam and amiodarone. Therefore, pharmaceutical compatibility of drug injection solutions should be checked before giving any intravenous drug mixtures.

Pharmacokinetic Interactions

Having gained entry into the body, drug interactions may occur when one drug alters the absorption, distribution, metabolism or excretion (ADME) of another drug given concurrently. Such pharmacokinetic interactions may result in either an increased or decreased amount of a drug available to exert its pharmacological effects in the body. Examples of drug interactions involving different ADME processes and the expected clinical outcomes are given in Table 10.1. It is important to realize that not all

Table 10.1 Types of pharmacokinetic drug interactions, with representative examples and the expected clinical outcomes

Pharmacokinetic process involved	Combined clinical effect	Example with beneficial outcome	Example with harmful outcome
Absorption: (a) Decreased	Reduced	Acetaminophen + activated charcoal (Activated charcoal can adsorb acetaminophen and prevent its GI absorption during an overdose)	Tetracycline + FeSO$_4$ (Fe^{2+} in iron supplements or food can form a complex with tetracycline and reduces the antibiotic absorption); Ketoconazole + antacids (Antacids, e.g., Mg(OH)$_2$, raise the pH of GI content, which reduce the dissolution of ketoconazole and hence its absorption)
	Enhanced	Lidocaine + epinephrine (Epinephrine causes vasoconstriction which slows the systemic absorption of the injected lidocaine, thereby prolonging the duration of the local anesthetic action and reduces its systemic toxicity risk)	
(b) Increased	Enhanced	Ketoconazole + Coca Cola (Coca Cola reduces the pH of GI content, which increases the dissolution of ketoconazole and hence its absorption)	
Distribution: (a) Displacement of plasma protein binding	Enhanced		Warfarin + sulfonamides (Sulfonamides, such as sulfamethoxazole, displace warfarin from its plasma protein bound site, which transiently increase its plasma concentration. This would normally be of little clinical significance if not for the simultaneous inhibition of its metabolism by the sulfonamide resulting in increased risk of hemorrhage)
(b) Displacement of tissue protein binding	Enhanced		Digoxin + quinidine (As in the example above, quinidine displaces digoxin from its tissue bound site and inhibits also its metabolism at the same time, resulting in digoxin overdose)

(continued)

Table 10.1 (continued)

Metabolism:			
(a) Enzyme induction	Reduced	Bilirubin + phenobarbital (Phenobarbital increases the conjugation and excretion of bilirubin; hence reduces the risk of hyperbilirubinemia in neonates)	Warfarin + rifampin (Rifampin accelerates the metabolism of warfarin and reduces its anti-coagulant effect)
	Enhanced		Acetaminophen + rifampin (Rifampin induces the formation of CYP450-mediated reactive metabolite of acetaminophen and potentiate its risk of hepatotoxicity)
(b) Enzyme inhibition	Enhanced		Phenytoin + erythromycin (Erythromycin inhibits the hepatic metabolism of phenytoin, resulting in its accumulation and this increases its risk of toxicity)
Excretion:			
(a) Changes in urinary pH	Reduced	Aspirin + NaHCO$_3$ (Sodium bicarbonate raises the pH of urine which reduces the tubular reabsorption of aspirin, a weak acid, and accelerates its urinary excretion during overdose [see Chap. 27])	
(b) Interference in tubular secretary mechanisms	Enhanced	Penicillin + probenecid (Probenecid inhibits the tubular secretion of penicillin, resulting in reduced excretion and prolonged duration of antibiotic action.	Methotrexate + aspirin (Aspirin inhibits the tubular secretion of methotrexate, resulting in reduced excretion and increased toxicity)

reduced combined clinical effects are necessarily harmful and not all enhanced combined clinical effects are necessarily beneficial drug interactions.

Pharmacodynamic Interaction

This occurs when the action or effect of a drug modulates that of another drug, producing a clinical response that differs from that anticipated from either drug when taken individually. In acute care, careful selection of drugs and meticulous monitoring maximize the treatment benefit and prevent untoward effect arising from the combination strategy. Not uncommonly, drugs given in combination intentionally for clinical benefits turn harmful either by over-amplifying a drug's effect or by reducing its efficacy. These interactions, beneficial or harmful, are summarized, with representative examples, in Table 10.2.

Table 10.2 Types of pharmacodynamic drug interactions, with representative examples of two drugs

Interaction	Combination effect (C) compared with summation of individual drug effect (A+B)	Example with beneficial outcome	Example with harmful outcome
Synergism	C > A+B	Aminoglycosides + penicillins (Penicillins are bactericidal through bacterial cell wall destruction, which also enhances aminoglycoside transport into cell and its bactericidal effect)	Barbiturates + opiates (Both are CNS depressant acting on different targets but give rise to similar/common effect that are thus augmented: sedation and respiratory depression)
Summation	C = A+B	Aspirin + acetaminophen (Acetaminophen lacks anti-inflammatory action but adds to the antipyretic and analgesic effect of aspirin)	Macrolides + quinolones (Both antibiotic groups have the potential of inducing heart arrhythmia)
Antagonism	C < A+B	Opiates + naloxone (Naloxone as the opioid receptor blocker reverses opiate effect in acute poisoning, i.e., respiratory depression. [see Chap. 7]) Copper + penicillamine (Penicillamine binds copper and reduces its harmful effect during copper poisoning)	Warfarin + vitamin K (Supplementary vitamin K disturbs the anticoagulation state maintained by warfarin, where the synthesis of vitamin K-dependent anticoagulants has been inhibited. This can lead to suboptimal or failure of anticoagulation therapy with a shortened INR)

The interaction can get more complicated with increasing number of drugs taken, as commonly occur in the practice of polypharmacy

Key Concepts

- Multiple drug prescription can cause drug-drug interactions that may produce beneficial or adverse outcomes.
- Drug interactions may also occur between drugs and other xenobiotics such as food constituents, environmental pollutants and herbal products.
- Drug interactions can be attributed to pharmaceutical, pharmacokinetic or pharmacodynamic factors.
- Drugs should only be prescribed in combination when necessary and clearly indicated for their therapeutic benefits.
- Where drug interactions may arise, the clinical condition of the patient should be carefully monitored for either therapeutic achievement or toxicity development.

Summary

Drug interactions include not only drug-drug interactions but also interactions between drugs and other xenobiotics such as food constituents, environmental pollutants and herbal products. Pharmacokinetic and pharmacodynamic interactions, as well as pharmaceutical incompatibility of drug mixtures, contribute to most of the clinically important drug interactions. The clinical outcome of intended therapy (intentional or unintentional combination of drugs) can be altered, producing beneficial or toxic effect. Careful selection of drugs for combined regimen and meticulous monitoring are important to ensure the therapy is optimized.

Further Reading

1. Boucher BA, Wood GC, Swanson JM. Pharmacokinetic changes in critical illness. Crit Care Clin. 2006;22:255–71.
2. British National Formulary (BNF) 59. London: BMJ Group and Pharmaceutical Press; March 2010. Appendix 1: Interactions; pp. 771–859.
3. Brunton L, Blumenthal D, Buxton I, Parker K, editors. Goodman & Gilman manual of pharmacology and therapeutics. 11th ed. New York: McGraw-Hill; 2008.
4. Hughes SG. Prescribing for the elderly patient: why do we need to exercise caution? Br J Clin Pharmacol. 1998;46:531–3.
5. Lynch T, Price A. The effect of cytochrome P450 metabolism on drug response, interactions, and adverse effects. Am Fam Physician. 2007;76:391–6.
6. Murney P. To mix or not to mix – compatibilities of parenteral drug solutions. Aust Prescr. 2008;31(4):98–101.

Part II
Pharmacology of the Various Body Systems

Chapter 11
Drugs and the Cardiovascular System

Noorjahan Haneem Md Hashim

Abstract The function of the cardiovascular system is to deliver oxygen and nutrients to the tissues. Failure of the cardiovascular system will result in tissue hypoxia. There are many categories of drugs that can be used to support the cardiovascular system by acting on the heart (pump) and/or the vessels (conduit). Many of these drugs act on the autonomic nervous system through various receptors and ion channels. Inotropes improve the contractility of the heart, whilst antiarrhythmic agents restore sinus rhythm and optimise heart rate for maximal efficiency of the pump. The oxygen delivery to the myocardium is maintained with the aid of coronary vasodilators in patients with ischemic heart disease. Vasopressors are administered to increase perfusion pressure and so improve flow, whilst vasodilators are used to treat hypertension which helps to decrease the work of the heart, while improving the blood flow to the myocardium. Most drugs administered to improve the function of the cardiovascular system have multiple mechanisms of action and may have more than one effect on the cardiovascular system, so they must be administered with caution and titrated to effect, and their effects monitored closely.

Keywords Cardiovascular support • Inotropes • Vasopressors • Vasodilators • Antiarrhythmics

Introduction

The function of the cardiovascular system is to deliver oxygen and energy sources to the tissues and to remove carbon dioxide and other waste products. The cardiovascular system is basically composed of the heart (the pump), the vessels (conduits) and blood (carriers of oxygen mainly combined with haemoglobin).

When any component in the system fails, oxygen delivery is jeopardised and carbon dioxide and waste products accumulate. There are many categories of drugs that can be used to optimise the function of the cardiovascular system, which act at

N.H. Md Hashim, M.B.B.S., M.Anaes (✉)
Department of Anesthesiology, Faculty of Medicine, University of Malaya,
50603 Kuala Lumpur, Malaysia
e-mail: noorjahan@um.edu.my

© Springer International Publishing Switzerland 2015
Y.K. Chan et al. (eds.), *Pharmacological Basis of Acute Care*,
DOI 10.1007/978-3-319-10386-0_11

different sites and through different mechanisms. For these drugs to function optimally, it is vital to look for, and treat reversible causes of cardiovascular compromise, e.g., tension pneumothorax, cardiac tamponade, hypoxemia, acidosis, hypovolemia and electrolyte imbalance, as these conditions will affect the patient's response to therapy.

Drugs Working on the Pump

The cardiac output from the heart must be adequate to ensure adequate perfusion. Cardiac output can be impaired by decrease in cardiac contractility, abnormalities in cardiac rhythm, decreases in preload (venous return) and increases in afterload (resistance).

Inotropes

Inotropes increase the cardiac output by improving pump function through increasing the strength of contraction, the rate of contraction and the rate of relaxation. The cardiac actin-myosin interaction is augmented by increased intracellular Ca^{2+}, so these drugs act to increase intracellular Ca^{2+}, via different routes:

1. **Amine sympathomimetic agents (catecholamines)**: when these drugs bind to the adrenergic receptors, they activate a cascade of changes inside the cell (Fig. 11.1). These changes increase intracellular Ca^{2+} via increasing intracellular cAMP concentrations. The strength of their actions depends on the dose dependent dominant receptor activation and adrenergic receptor density in cardiac tissues. Examples of such drugs include dopamine, dobutamine and adrenaline. Due to their potency, they are administered via continuous intravenous infusion. They have rapid onset and offset. Offset is mostly due to neuronal re-uptake. They are metabolised by the enzymes mono-amine oxidase (MAO) and cathecol-O-methyl transferase (COMT).

   Dopamine produces different effects at different doses. At low doses it acts mainly on dopaminergic receptors, giving rise to positive inotropy. β-adrenergic effects (positive inotropy and chronotropy) followed by α effects (increased systemic vascular resistance (SVR)) are seen as the dose increases, with predominantly α effects at high doses causing marked increase in SVR.

   Dobutamine is a synthetic sympathomimetic. It is also administered via intravenous infusion. It mainly affects β-adrenoceptors, causing positive inotropy, and has variable effects on vascular resistance and heart rate.

   Adrenaline has both α and β adrenergic effects, causing positive inotropy, chronotropy and increased SVR.

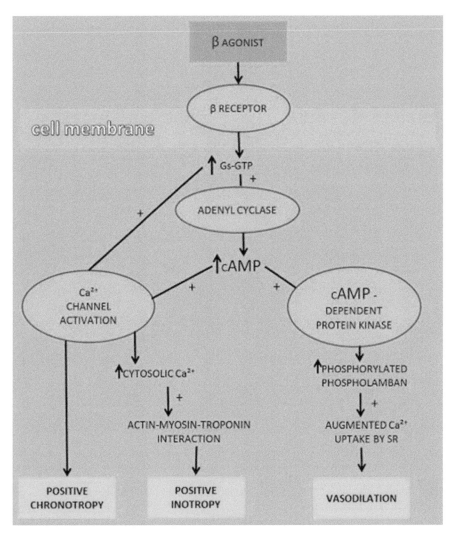

Fig. 11.1 The mechanism by which catecholamine exerts its effects. Gs-GTP: Stimulating Guanosine nucleotide-binding protein-Guanosine Triphosphate complex; cAMP: cyclic Adenosine Monophosphate; Ca^{2+}: Calcium; SR: sarcoplasmic reticulum (Reproduced with permission from Wolters Kluwer Health; Circulation 2008; 118: 1047–56)

2. **Phosphodiesterase-III inhibitors (PDIs)**: these drugs work inside the cell to prevent the breakdown of cAMP, thereby increasing intracellular Ca^{2+} concentrations (Fig. 11.2). A commonly used drug in this category is milrinone.

   Milrinone is also administered via intravenous infusion. It is also called an inodilator, as it has positive inotropic effects while giving rise to reduced SVR. It is metabolised in the liver.

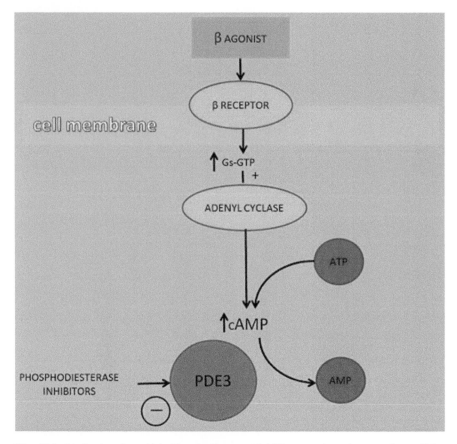

Fig. 11.2 Mechanism by which Phosphodiesterase inhibitors work to increase intracellular calcium concentrations. Gs-GTP: Stimulating Guanosine nucleotide-binding protein-Guanosine triphosphate complex; cAMP: cyclic adenosine monophosphate; ATP: Adenosine triphosphate; PDE3: Phosphodiesterase 3 (Reproduced with permission from Wolters Kluwer Health; Circulation 2008; 118: 1047–56)

3. **Digitalis**: This glycoside inhibits Na^+/K^+-ATPase thus increasing the intracellular sodium concentrations. This in turn increases intracellular calcium concentrations (via the Na^+/Ca^{2+} exchanger), augmenting the actin-myosin interaction and giving rise to stronger contractions. This also increases the action potential (AP) duration and the refractory period of the atrioventricular (AV) node and the bundle of His, thereby reducing AV conductivity and reducing ventricular rate, making it useful in the treatment of arrhythmias. A common digitalis is digoxin.

Coronary Vasodilators

Pump dysfunction may be due to myocardial ischemia caused by oxygen supply–demand imbalance. Caregivers attempt to maintain blood supply to the myocardium and reduce its oxygen requirement to ensure adequate oxygen supply. This in turn will improve its pump function. One method used to maintain the myocardial oxygen demand and supply balance is by administering coronary vasodilators, i.e., nitrates.

Glyceryl trinitrate (GTN) is a direct acting vasodilator that is metabolised to nitric oxide (nitric oxide donor). Activation of M_3 receptors in blood vessels increases nitric oxide synthesis which then diffuses into smooth muscle cells and increases cyclic GMP. This acts on myosin light chain phosphatase that causes inactivation of myosin light chain. The resulting smooth muscle relaxation causes vasodilation. Dilatation of epicardial and interconnecting coronary arteries increases coronary blood flow both in normal and stenotic vessels. At low doses, the effect of GTN is seen in capacitance vessels (veins), affecting preload (venous return) more than afterload (SVR). This reduces left ventricular end diastolic pressures (LVEDP) and myocardial wall tension, increasing sub-endocardial blood flow. Care must be taken, however, as at high doses GTN causes a reduction in the SVR as well. This will in turn reduce coronary perfusion pressure (aortic pressure).

GTN is indicated in the treatment of myocardial ischemia (during acute coronary syndrome), hypertensive crises, heart failure and coronary vasospasm. It is only suitable for titrated intravenous infusions (0.1–10 mcg/kg/min), sublingual delivery (300–500 mcg) and transdermal application (5–10 mg/24 h). It has immediate onset and offset. $t_{1/2\beta}$ is 1–4 min. It is mostly metabolised in the liver to both active and inactive metabolites. Adverse reactions include tolerance, rebound hypertension, increased intrapulmonary shunt, loss of hypoxic pulmonary vasoconstriction and methemoglobinemia. Symptoms include tachycardia, dizziness, drowsiness, vertigo, facial flushing, weakness and fainting.

Antiarrhythmia Drugs

The optimal rhythm for maximal efficiency of pump function is sinus rhythm. In this state, there is synchronized and timely flow of blood from the atrium to the ventricles, which is lost in the presence of arrhythmias. Caregivers attempt to control cardiac rhythm, failing which they try to control the rate so that there will still be adequate filling to ensure reasonable output, e.g., in atrial fibrillation, rate control is often used.

Drugs that maintain sinus rhythm or control the heart rate are called antiarrhythmia agents. A popular classification of these drugs is the Vaughn Williams classification. Class I and IV acts on ion channels, Class II on receptors and Class III acts on both. Most of these drugs have mechanisms of action which overlap.

Class I Antiarrhythmic Drugs

These are sodium channel blockers. They are further divided into three groups.

Class Ia prolongs action potential duration and reduces AV conduction rate. An example is procainamide, used in the treatment of ventricular arrhythmia and atrial fibrillation (AF), administered intravenously or orally. It is metabolized by the liver, with an active metabolite (NAPA, a potassium channel blocker). Its $t_{1/2}$ is 3–5 h. Common side effects include ventricular arrhythmias, including torsades de pointes, lupus like syndromes and blood dyscrasias.

Class Ib agents reduce action potential duration, but have no effect on AV conduction rate. An example is lignocaine, which is indicated in ventricular arrhythmias. Its route of administration is intravenous and its $t_{1/2}$ is 1.5–2 h. Common side effects include tremors, paraesthesia and seizures.

Class Ic agents reduce AV conduction rate and have no effect on action potential duration. An example is flecainide, which is indicated in the treatment of paroxysmal atrial fibrillation and Wolf-Parkinson-White syndrome. Its route of administration is oral. It undergoes hepatic metabolism and its $t_{1/2}$ is 20 h. It is contraindicated in patients with structural heart disease, and side effects include arrhythmias (especially in post-myocardial infarct patients), heart failure, headache and blurred vision.

Class II Antiarrhythmic Drugs

These are β-receptor antagonists. They bind to β_1-adrenoceptors located in cardiac nodal tissue, the conducting system, and contracting myocytes, inhibiting the effects of adrenaline on the β_1-adrenoceptors. Blockade of β-adrenoceptors inhibits activation of calcium channels, leading to a reduction of AV node conduction. One example is esmolol, which is used in the treatment of paroxysmal supraventricular tachycardia, rate control in non-pre-excited AF or atrial flutter, ectopic atrial tachycardia, inappropriate sinus tachycardia, polymorphic ventricular tachycardia (VT) due to torsades de pointes or myocardial ischemia. It is administered intravenously.

Class III Antiarrhythmic Drugs

These are potassium channel blockers. Agents in this group prolong cardiac action potential and refractory period. A popular example is amiodarone, used in the treatment of paroxysmal atrial, nodal and ventricular tachycardias (including atrial fibrillation and persistent ventricular tachycardia). It may be administered intravenously (5 mg/kg (300 mg over 60 min), then 15 mg/kg (900 mg over 23 h)), or orally (600 mg/day, then reduced to 200 mg/day after two weeks). Known side effects are sinus bradycardia, thyroid dysfunction and pulmonary fibrosis.

Class IV Antiarrhythmic Drugs

These drugs are calcium channel blockers. Agents in this group reduce action potential duration and AV conduction rate. An example is verapamil, indicated in the treatment of supraventricular tachycardia (SVT). It is administered intravenously in boluses of 5–10 mg (0.075–0.15 mg/kg) over 2 min, repeated after 10 min if SVT persists.

Adenosine

Adenosine is an endogenous nucleoside found in all the cells of the body, which is used specifically in the treatment of superventricular tachycardia. Adenosine slows conduction through the atrioventricular node but its mechanism of action is unclear. The effect of adenosine is terminated by cellular uptake, principally by erythrocytes and endothelial cells, so it has a very short half-life of less than 10 s. It is administered by a rapid intravenous bolus of 6 mg, followed by rapid saline flush. If the initial dose is ineffective after 1–2 min, a second 12 mg bolus may be given which can be repeated once more, if necessary. Potentially life threatening heart block may occur with adenosine administration (usually self-limiting because of the short duration of action), so resuscitation facilities must be available. Ventricular fibrillation may rarely develop if adenosine is given in the presence of digoxin or verapamil. Other side effects include flushing, headache, bronchospasm and hypotension.

Drugs Working on the Conduit

Flow in a vessel is directly proportional to pressure and to the fourth power of radius. This is demonstrated in the Hagen-Poiseuille equation where:

$$Q = \frac{\delta P \pi r^4}{8 \eta l}$$

Where Q is flow, δP is pressure difference or perfusion pressure, r is the radius, η the viscosity and l the length of the conduit.

So, blood flow in the vessels can be controlled by administering vasopressors or vasodilators.

Vasopressors

Flow may be decreased at low perfusion pressures, due to the relationship mentioned above. In an attempt to increase flow by improving the perfusion pressure, vasopressors can be administered, particularly in hypotensive states due to vasodilatation, such as in septicemic shock. The increase in SVR brought about by vasopressors allows perfusion of the vital organs, i.e., the brain and heart, at the expense of other organs such as the gastrointestinal tract and skin. These agents cause vasoconstriction mainly in the arterial system through increased intracellular Ca^{2+} concentration. Two groups of agents can be used:

1. **Sympathomimetic agents (catecholamines)**: When these drugs bind to the G-protein coupled α_1-adrenergic receptors, it activates a cascade of changes inside the cell. These changes increase intracellular Ca^{2+}, leading to vasoconstriction. This concept is illustrated in Fig. 11.3. Examples of agents used as vasopressors are directly acting norepinephrine and phenylephrine, indirectly acting ephedrine and dose-dependent dopamine and epinephrine.

   Ephedrine causes vasoconstriction via three mechanisms: by directly stimulating α_1 and β-adrenoreceptors and by stimulating the release of norepinephrine from presynaptic nerve endings. It is administered intravenously in aliquots of 6 mg. It has rapid onset and offset. Tachyphylaxis may occur with repeated administrations. It is excreted unchanged by the kidneys.

2. **Vasopressin**: Intracellular Ca^{2+} is also increased when vasopressin (a hormone secreted by the pituitary) binds to the G-protein coupled V_1 receptors found on vascular smooth muscle of the systemic, splanchnic, renal, and coronary circulations. The events are similar to the events seen when α_1-adrenergic receptors are stimulated (Fig. 11.3).

Vasodilators

Although vasodilators can improve blood flow such as in the case of glyceryl trinitrate (GTN) improving coronary perfusion, vasodilators are mainly used to treat undesirable hypertension rather than for improvement of blood flow per se. Reduction in the systemic vascular resistance (SVR) or afterload can improve the cardiac output in the presence of hypertensive heart failure and improves the myocardial oxygen supply/demand balance by decreasing the work done by the heart.

Sodium nitroprusside (SNP) is a nitric oxide donor, causing smooth muscle relaxation and vasodilation of both arterioles and veins. It causes a drop in blood pressure by reducing SVR at low and high doses, and increased venous capacitance at high doses. Although it has no direct effects on the heart, reflex tachycardia may occur.

SNP is indicated in the treatment of perioperative increase in mean arterial pressure (MAP) secondary to increased SVR, in hypotensive anaesthesia and for no-flow syndrome post-angioplasty. The dose is 0.25–10 mcg/kg/min.

SNP should only be administered via intravenous infusion. It is rapidly metabolized to cyanide and then to thiocyanate, which is eliminated by the kidneys. Accumulation occurs in patients with renal failure.

It is contraindicated in patients with hypothyroidism and renal failure. Adverse reactions are similar to GTN. It can also cause coronary steal syndrome and cyanide and thiocyanate toxicity.

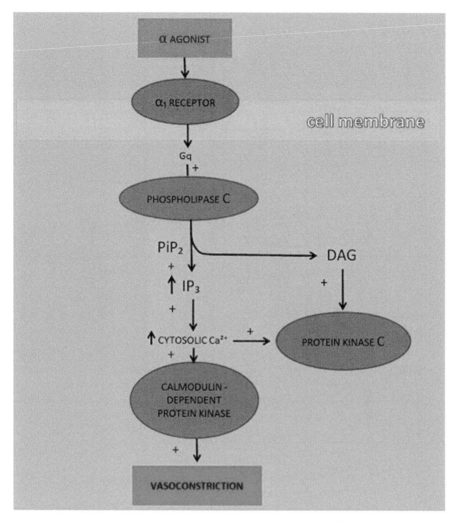

Fig. 11.3 The mechanism of vasoconstriction due to α_1-receptor stimulation. Gq: stimulating guanosine nucleotide-binding protein, PIP2: a membrane phosphoinositol; IP3: Inositol triphosphate; DAG: Diacylglycerol (Reproduced with permission from Wolters Kluwer Health; Circulation 2008; 118:1047–56)

Monitoring

It is important to remember that these agents have multiple functions, and may have more than one effect on the cardiovascular system, depending on the body's internal compensatory mechanisms, the receptor dominance of the organ system, dose of the agents used and interactions with other drugs.

The choice of agents should be selective, and patients should be monitored for their response to the drugs administered and for the presence of toxicity. This requires multiple target endpoints:

Hemodynamic Monitors

Heart rate, blood pressure, pulse pressure and central venous pressure are definite hemodynamic parameters of the cardiovascular system we can use to guide therapy.

Surrogate Markers of Regional Tissue Perfusion

There are indirect parameters of adequacy of regional tissue perfusion in the form of mentation, urine output, capillary refill time and feeding tolerance.

Global Tissue Perfusion

As mentioned previously the role of the cardiovascular system is to deliver oxygen to the tissues. How well the system is fulfilling this role can be monitored by determination of global tissue perfusion through the serum lactate levels, base excess or mixed venous oxygen tension.

Blood Levels of Drugs

Some drugs have small therapeutic ranges or long half-lives, and so the blood levels of these drugs need to be monitored. A good example of this is digoxin, which has a half-life of 36 h, especially in patients who have renal failure.

Administration of Drugs in Acute Care

Most of these drugs are very potent, and some have short half-lives. The safest way to administer these drugs is via a continuous central venous infusion, as this allows titration of the drugs to the patient's response. As most vasopressors cause peripheral vasoconstriction, the central venous route allows a more reliable delivery of these agents, while reducing the risk of local tissue necrosis in case of extravasation.

Key Concepts

- In the management of patients with impaired oxygen delivery due to failure of the cardiovascular system, the presence of conditions such as tension pneumothorax, cardiac tamponade, hypoxemia, acidosis, hypovolemia and electrolyte imbalance must be excluded and/or treated to allow drugs acting on the heart and vessels to work.
- Agents used to support the cardiovascular system have multiple mechanisms of action, and may have more than one effect on the cardiovascular system.
- The choice of agents should be selective, and patients should be monitored continuously with multiple target endpoints such as hemodynamic parameters, and markers of tissue perfusion.
- The safest way to administer these agents is via central venous access, titrating to the patient's response.

Summary

Failure of the cardiovascular system jeopardises oxygen delivery. It is crucial that this system is supported in the acute care setting to prevent tissue hypoxia. There are many categories of drugs that can be used to optimize the cardiovascular system. Many of these drugs cause changes in the autonomic nervous system through various receptors (adrenergic, dopaminergic, vasopressin) and ion channels (calcium, sodium and potassium). Positive inotropic drugs improve the strength of cardiac contractions while antiarrhythmics maintain sinus rhythm and optimal heart rate to ensure the maximal efficiency of the pump. The blood supply to the myocardium, and hence its oxygen supply is improved by coronary vasodilators. The rationale for administering drugs that work on the conduit is explained by the Hagen-Poiseuille equation, where flow is directly proportional to perfusion pressure and the radius of the vessels. Vasopressors are administered to increase perfusion pressure and vasodilators improve myocardial function by decreasing the afterload when used to treat hypertension. These drugs have multiple mechanisms of action and may have more than one effect on the cardiovascular system. It is extremely important that they are administered with caution and their effects and side effects are monitored closely.

Further Reading

1. Benham-Hermetz J, Lambert M, Stephens RCM. Cardiovascular failure, inotropes and vasopressors. Br J Hosp Med. 2012;73:C74–7.
2. Dougherty M, Webb ST. Inotropes and vasoactives. In: Mackay JH, Arrowsmith JE, editors. Core topics in cardiac anaesthesia. 2nd ed. Cambridge: Cambridge University Press; 2012.

3. Ellender TJ, Skinner JC. The use of vasopressors and inotropes in the emergency medical treatment of shock. Emerg Med Clin North Am. 2008;28:759–86.
4. Homoud MK. Introduction to antiarrhythmic agents. Tufts-New England Medical Center. 2008. Cited May 2014. Available from http://ocw.tufts.edu/data/50/636944.pdf. Last accessed 22nd June 2014.
5. Klabunde RE. Cardiovascular pharmacology concepts. Available from http://www.cvpharmacology.com. Last accessed 12th June 2014.
6. Overgaard CB, Dzavik V. Inotropes and vasopressors: review of physiology and clinical use in cardiovascular disease. Circulation. 2008;118:1047–56.
7. Tristan Walker. Basics of cardiac pharmacology. Available from http://learnpediatrics.com/body-systems/cardiology/basics-of-cardiac-pharmacology/. Last accessed 22nd June 2014.
8. Vaughn Williams EM. A classification of antiarrhythmic actions reassessed after a decade of new drugs. J Clin Pharmacol. 1984;24(4):129–47.

Chapter 12
Drugs and the Respiratory System

Yong-Kek Pang

Abstract The respiratory system serves as an important interface for gas exchange. It is vital to maintain the health of the airways to ensure a free flow of gases. Certain diseases that affect the airways and the gas exchange site are best treated with drugs delivered directly to the target sites via the inhalational route. This will not only result in immediate action but will keep the required drug dosages to the minimum and thus reduce their systemic side-effects. The non-inhalational mode is an alternative option when the passage to the gas exchange interface is too severely obstructed or if the drugs can only be given via the systemic route. There are two major classes of bronchodilators delivered via the inhalational mode in acute care, i.e., the beta 2 agonists and the anticholinergics. The two differ in both the onset of action as well as the duration of action. A combination of drugs from different classes is often employed in the acute situation to achieve the best bronchodilatory effect. Drug delivery for patients with poor inhalational technique can be assisted by means of a spacer or a nebulizer.

Keywords Aerosol-holding chamber • Anti-inflammatory agents • Corticosteroids • Bronchodilators • Anticholinergics • Beta 2 agonist • Gas exchange • Inhalational mode • Mechanism of action • Methylxanthines • Aminophylline • Theophylline • Nebulizer • Jet nebulizer • Ultrasonic nebulizer • Non-inhalational mode • Respiratory system • Spacer

Introduction

The respiratory system serves as one of the interfaces of the body with the environment. It is the system that allows the entrance of the most important element for survival, i.e., oxygen, and the exit of the waste gas carbon dioxide. The health of this system is essential and there is a wide array of drugs which play vital roles in

Y.-K. Pang, M.D., MRCP (✉)
Division of Respiratory Medicine, Department of Medicine, Faculty of Medicine,
University of Malaya, 50603 Kuala Lumpur, Malaysia
e-mail: ykpang@ummc.edu.my

© Springer International Publishing Switzerland 2015
Y.K. Chan et al. (eds.), *Pharmacological Basis of Acute Care*,
DOI 10.1007/978-3-319-10386-0_12

sustaining the patency of the small airways in certain illnesses. Most of these are best administered via the respiratory system itself; while others are given through non-inhalational routes.

Pathophysiology

The respiratory system is a unique system in the human body. It consists of the conduit or airway, the gas exchange site and the respiratory control center. The total gas exchange surface of the adult lungs is equal to 70–100 m^2 (or half the size of a tennis court). This gas exchange function is facilitated by the circulatory system, whereby blood is brought to close proximity with the alveolar epithelia. Every minute in a 70 kg man, about 5 L of air is breathed in and brought to the gas exchange site.

Patency of the airways is of crucial importance in order for adequate gas exchanges to take place. Any impediment to this increases the work of breathing and energy required in this process, which in normal circumstances only takes up 1–2 % of the body's daily energy consumption.

The unique structure of the airways and alveoli provide a good opportunity for the delivery of drugs via the inhalation mode straight to the intended sites of action, i.e. the epithelium, the bronchial smooth muscles or the pulmonary vessels.

Inhalation Mode of Drug Delivery

This mode of delivery offers some potential advantages compared to the oral route:

1. The drug is delivered directly to the target site, thus allowing immediate action.
2. The drug bypasses the first-pass effect of the liver and does not need to be distributed to the rest of the body - hence, the dose required is generally quite low.
3. With a relatively low dosage, the systemic effects of inhaled drugs are significantly reduced.

However, the effectiveness of drugs delivered via the inhalation route is very much dependent on the technique of inhalation, the drug particle size, the solvent and the design of the inhaler. In the normal lungs, the drug after inhalation is quite evenly distributed. However, in the pathological lungs, the drug distribution may be uneven due to one or more of the following factors: bronchoconstriction, airway inflammation, retention of secretion, endobronchial lesion, emphysema and fibrosis of the lungs.

In acute situations, the coordination and synchronization of the patients with the inhaler may be jeopardized due to tachypnea, restlessness, drowsiness or confusion. The utilization of a spacer (also known as aerosol-holding chamber) which obviates the need for synchronization and slows down the delivery speed of aerosols from a pressurized metered-dose inhaler, helps to overcome this problem. Alternatively, the drug could be administered via a nebulizer. There are different types of nebulizers, the jet nebulizer, ultrasonic wave nebulizer and vibrating mesh nebulizer.

Table 12.1 Inhaled β_2 agonists

Categories	Drugs	Onset of action	Duration of action
Short-acting β_2 agonists (SABA)	Salbutamol Fenoterol Terbutaline	5–10 min	4–6 h
Long-acting β_2 agonists (LABA)	Formoterol	5–10 min	12 h
	Salmeterol	20 min	
Ultra-long-acting β_2 agonists (LABA)	Indacaterol	5–10 min	24 h

For patients with severe airway obstruction secondary to bronchoconstriction or retention of secretions, a higher dose of bronchodilator is required to achieve similar treatment effects. Most providers also opt for combination therapy with different classes of drugs. As they work synergistically through different pathways, lower doses are usually required to achieve the same therapeutic effects, thereby reducing the systemic side effects. In the most severe cases, systemic administration of the drug should be considered.

Inhaled Beta-2 Agonists

Inhaled β_2 agonists are used for managing bronchospasm. They are classified according to their duration of action (Table 12.1). Most have rapid onset of action of about 5–10 min. The long-acting β_2 agonists are more useful as maintenance therapy for patients with persistent symptoms.

Mechanism of Action of β_2 Agonists

Activation of β_2 receptors (β_2AR) by β_2 agonist results in activation of adenylyl cyclase (AC) via a stimulatory G protein (Gs). This leads to an increase in intracellular cyclic AMP (cAMP) and activation of phosphokinase A (PKA) (Fig. 12.1).

PKA phosphorylates a variety of target substrates, the activities of which cause among others, reduction of myosin light chain kinase (MLCK) activity and intracellular calcium. In addition, β_2 agonists open up calcium-activated potassium conductance channels and cause hyperpolarization of airway smooth muscle cells. The combination of decreased MLCK activity, decreased intracellular calcium, and increased membrane potassium conductance lead to smooth muscle relaxation and bronchodilatation.

Regular use of β_2 agonists, particularly the SABAs, may result in down regulation of β_2 receptors and reduced efficacy. Consequently, for these individuals, higher doses are required during acute exacerbation of symptoms. Their side effects include palpitation, tachycardia, nervousness, tremor and arrhythmia.

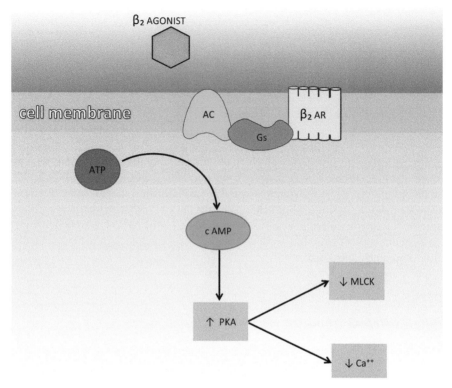

Fig. 12.1 Activation of adenylyl cyclase (AC) leads to an increase in intracellular cyclic AMP and activation of phosphokinase A (PKA). PKA phosphorylates a variety of target substrates which result in smooth muscle relaxation and bronchodilatation. Note: β_2AR – Beta-2 Agonist Receptor; Gs – Stimulatory G protein; MLCK – Myosin Light Chain Kinase

Table 12.2 Inhaled anticholinergics

Categories	Drugs	Onset of action	Duration of action
Short-acting anticholinergics (SAAC)	Ipratropium bromide Oxitropium bromide	15–20 min	6–8 h
Long-acting anticholinergics (LAAC)	Tiotropium bromide	20–30 min	≥ 24 h or more
	Glycopyrronium bromide	5–15 min	24 h
	Aclidinium bromide	15–20 min	12 h

Inhaled Anticholinergics

This category of drugs is also used for managing patients with bronchospasm. Likewise, they are classified according to their duration of action (Table 12.2). The onset of action of ipratropium is slower (15–20 min) than the β_2 agonists and hence it is not ideal as a reliever drug. However, it could be used in combination with short-acting β_2 agonists to relieve acute symptoms. Combination therapy of long-acting anticholinergics and long-acting β_2 agonists are being used as maintenance therapy for patients with chronic obstructive pulmonary disease (COPD), who do not achieve optimal symptom control with either drug alone.

Mechanism of Action of Anticholinergics

Under normal resting condition, the airways are maintained in tonic bronchoconstriction due to continuous background activities of parasympathetic nerves. This effect is exerted through acetylcholine released from the preganglionic and postganglionic nerve endings and mediated via the M_1 receptors (parasympathetic ganglia) and M_3 receptors (the smooth muscles of the bronchi). The action of acetylcholine can be blocked at the M_1 and M_3 receptors by means of an anticholinergic (Fig. 12.2).

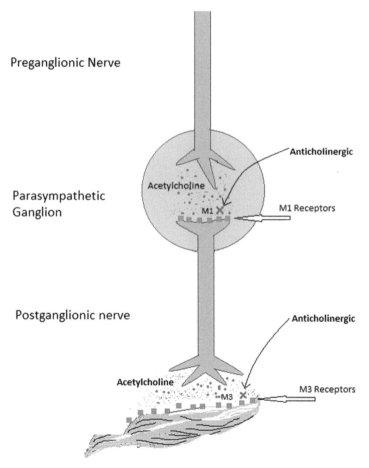

Fig. 12.2 Bronchoconstriction is caused by acetylcholine released from the preganglionic and postganglionic nerve endings and mediated via the M_1 receptors (at parasympathetic ganglia) and M_3 receptors (at smooth muscles of the bronchi). An anticholinergic agent reverses this effect

Side Effects

Their side effects include dry mouth, blurred vision, worsening of glaucoma and urinary retention.

Other Drugs Acting on the Respiratory System

Whilst many of these drugs may be administered by the inhalation route to the respiratory system, certain drugs are administered by the intravenous route in the acute situation, to produce effects in the respiratory system.

Methylxanthines

These drugs are used mainly for bronchodilatation. However, they have also been shown to have mild anti-inflammatory effect and inotropic effect on the myocardium and diaphragmatic muscles. A synergistic effect with corticosteroids has recently been demonstrated in several studies.

The two commonly used agents are theophylline (oral form) and aminophylline (intravenous form). As these drugs have narrow therapeutic windows, care should be taken with their use. Even when these drugs are within the therapeutic range, toxicity can occur. Close monitoring of blood levels is strongly advised. The side effects of these drugs include cardiac arrhythmias, seizure, headache, diuresis, nausea, epigastric/abdominal pain and diarrhoea.

Mechanism of Action

These drugs act through the inhibition of phosphodiesterase and this results in an increase in cAMP which leads to the activation of protein kinase A and other target proteins that culminate in the relaxation of airway smooth muscle. It also activates histone deacetylase which causes deactivation of core histones and thereby suppression of inflammatory genes and proteins.

Corticosteroids

This group of drugs is one of the commonest classes of drugs used in inflammatory lung diseases due to their potent anti-inflammatory effects. They are particularly useful in suppressing eosinophilic inflammation (e.g., asthma), but much less effective against neutrophilic inflammation (e.g., COPD).

In acute exacerbation of asthma, the oral and parenteral administrations of corticosteroids are equally effective. The onset of action for their anti-inflammatory effects is often delayed (about 6 h). Hence, they are administered with β_2 agonists or anticholinergics which have more immediate onset of action, while awaiting the anti-inflammatory effects to take place.

Mechanism of Action

The anti-inflammatory effects are mediated by direct binding of the glucocorticoid-glucocorticoid receptor complex to glucocorticoid responsive elements in the promoter region of gene. This glucocorticoid-glucocorticoid receptor complex also reduces inflammation through interaction with other transcription factors, in particular activating protein-1 or nuclear factor-$\kappa\beta$. Other anti-inflammatory effects are exerted through inhibition of many inflammation-associated molecules such as cytokines, chemokines, arachidonic acid metabolites, and adhesion molecules.

Side-Effects

The short-term use is generally safe. In the acute setting, the side effects include increased acid secretion with risk of gastrointestinal bleeding, increased risk of infection, impaired glucose tolerance, increased fluid retention and raised blood pressure, and sudden onset of confusion and insomnia.

Magnesium Sulfate

Intravenous magnesium sulfate has been shown to improve pulmonary function in stable asthmatic patients. In the acute setting, the benefits (in terms of lung function improvement and admission rate) of intravenous 2 g magnesium sulfate base at the emergency unit remain controversial. It appears that only those with severe airflow obstruction (FEV1 < 25 % predicted) may benefit from this treatment.

Leukotriene Inhibitors

Drugs in this class can be divided into two categories: the leukotriene receptor inhibitors (e.g., montelukast and zafirlukast) and 5-lipoxygenase inhibitor (e.g., zileuton). They are useful in managing stable asthma. In a double-blind, crossover study, intravenous montelukast had been shown to increase the pulmonary function and reduce admission rate compared to the oral administration. However, drugs in this class are yet to be established as standard agents in the management of acute asthma.

Expectorants

In inflammatory airway disorders and acute flare of airway inflammation from whatever cause, bronchial secretion may increase. This causes obstruction to the airflow and impediment of gas exchange. Expectoration of this secretion may help to relieve this obstruction. However, some patients have difficulty in bringing up this secretion (sputum).

Agents believed to have expectorant properties, such as guaifenesin and iodinated glycerol are often included in cough mixtures. However, there is no evidence that any of these agents works.

Mucolytics

Systematic review of clinical trials confirms the efficacy of oral N-acetylcysteine in COPD patients. However, due to its relatively high cost and the modest effect of this drug, its cost-effectiveness is questionable.

Drugs with Negative Effect on the Respiratory System

Drugs used in the management of other diseases may also impact on the respiratory system. Beta antagonists may cause bronchoconstriction and so it is important to avoid this class of drug in asthmatic patients or choose a more selective β_1 antagonist. Sedatives and narcotics may suppress the respiratory center resulting in hypoxemia and carbon dioxide retention. These include antihistamines, benzodiazepines, opioids and their derivatives.

Key Concepts

- Inhalational drugs are very useful to manage respiratory illnesses as they are delivered straight to the targets within the lungs in relatively low doses with reduced systemic side effects.
- Drugs with quick onset of action should be chosen to relieve acute symptoms.
- In patients with persistent symptoms, long-acting agents are more ideal to alleviate them.
- Combination of various drugs which act on different targets are more effective in improving the pulmonary function and should be considered in patients with severe symptoms.
- Certain drugs used in the management of other illnesses may impact the function of the respiratory system and hence caution must be exercised when using them.

Summary

In acute illness, increased metabolism is accompanied by an increased demand for gas exchange. This may be compromised in patients with airflow obstruction. Inhalational drugs are particularly effective in circumventing this limitation when caused by reversible bronchospasm. Systemic therapy is required when the drugs are only available in the oral/parenteral form or when the airways are too severely obstructed. It is also appropriate when suppression of systemic inflammation is necessary. A strategy of combining drugs acting via different pathways will produce the greatest desirable effect.

Further Reading

1. Barnes PJ. New therapies of COPD. Thorax. 1998;53:137–47.
2. Barnes PJ. Theophylline. Am J Respir Crit Care Med. 2003;167(6):813–8.
3. Barnes PJ. New drugs for asthma. Nat Rev Drug Discov. 2004;3:831–44.
4. Dockhorn RJ, Baumgartner RA, Leff JA, Noonand M, Vandormaela K, Strickere W. Comparison of the effects of intravenous and oral Montelukast on airway function: a double-blind, placebo controlled, three period, crossover study in asthmatic patients. Thorax. 2000;55:260–5.
5. Ferguson GT. Update on pharmacologic therapy for chronic obstructive pulmonary disease. Clin Chest Med. 2000;21:723–38.
6. Jonsson S, Kjartansson G, Gislason D, Helgason H. Comparison of the oral and intravenous routes for treating asthma with methylprednisolone and theophylline. Chest. 1988;94:723–6.
7. Poole PJ, Black PN. Oral mucolytic drugs for exacerbations of chronic obstructive pulmonary disease: systematic review. BMJ. 2001;322:1271–4.
8. Ratto D, Alfaro C, Sipsey J, Glovsky MM, Sharma OP. Are intravenous corticosteroids required in status asthmaticus? JAMA. 1988;260(4):527–9.
9. Rodrigo G, Rodrigo C, Burschtin O. Efficacy of magnesium sulfate in acute adult asthma: a meta-analysis of randomized trials. Am J Emerg Med. 2000;18:216–21.
10. Rolla G, Bucca C, Caria E, Arossa W, Bugiani M, Cesano L, Caropreso A. Acute effect of intravenous magnesium sulfate on airway obstruction of asthmatic patients. Ann Allergy. 1988;61:388–91.
11. Rowe BH, Bretzlaff J, Bourdon C, Bota G, Blitz S, Camargo CA. Magnesium sulfate for treating exacerbations of acute asthma in the emergency department. Cochrane Database of Systematic Reviews 2000, Issue 1. Art. No.: CD001490. DOI: 10.1002/14651858.CD001490
12. Rubin BK, Ramirez O, Ohar JA. Iodinated glycerol has no effect on pulmonary function, symptom score, or sputum properties in patients with stable chronic bronchitis. Chest. 1996;109:348–52.
13. Silverman RA, Osborn H, Runge J, Gallagher EJ, Chiang W, Feldman J, et al. IV magnesium sulfate in the treatment of acute severe asthma: a multicenter randomized controlled trial. Chest. 2002;122(2):489–97.
14. van der Velden VH. Glucocorticoids: mechanisms of action and anti-inflammatory potential in asthma. Mediators Inflamm. 1998;7(4):229–37.

Chapter 13
Drugs and the Central Nervous System

Mohd Shahnaz Hasan

Abstract As in all other organ systems, prevention of hypoxia is the aim in the management of central nervous system pathology, to prevent or minimize secondary brain damage and neuronal dysfunction. In the normal brain, auto-regulation, oxygen and carbon dioxide levels are the main determinants of cerebral blood flow. Once the physiological regulatory conditions have been breached, such as in trauma, bleeding or infection, the intracranial pressure rises and caregivers intervene to maintain adequate oxygenation and ventilation, and try to ensure that cerebral blood flow (and thus oxygen delivery) is adequate to meet the metabolic demand of the brain. This is more crucial in the brain than in any other organ in the body because its high metabolic requirements render the brain extremely sensitive to hypoxic conditions. The drugs used are mainly to modulate cerebral metabolism, cerebral perfusion pressure, cerebral blood flow and brain volume (cerebral edema). Adequate knowledge of cerebral physiology and the pharmacological effects of these drugs is vital to provide optimum care in patients with cerebral pathology.

Keywords Intracranial pressure • Cerebral blood flow • Cerebral metabolism • Cerebral perfusion pressure • Brain damage • Neuronal dysfunction

Introduction

Due to the high metabolic demand of the brain (20 % of total body oxygen consumption at rest), it is extremely sensitive to hypoxic conditions. Besides being the center of control of our voluntary functions, it is also the seat of our consciousness, intellect and personality; so any damage to our brain could result in a loss of physical function as well as a loss of 'self'. Thus, safeguarding the integrity of the brain is the prime objective of medical practice.

Brain tissue, cerebrospinal fluid (CSF) and blood are the main components within a rigid skull of fixed volume. Any increase in any of the components, unless accompanied by a corresponding decrease in one or more of the other components,

M.S. Hasan, M.B.B.S., M.Anaes (✉)
Department of Anesthesiology, Faculty of Medicine, University of Malaya,
50603 Kuala Lumpur, Malaysia
e-mail: shahnaz@ummc.edu.my

© Springer International Publishing Switzerland 2015   111
Y.K. Chan et al. (eds.), *Pharmacological Basis of Acute Care*,
DOI 10.1007/978-3-319-10386-0_13

leads to an increase in intracranial pressure, which eventually leads to a decrease in cerebral blood flow. Pharmacological agents that can influence these three components play a vital role, either directly or indirectly, in ensuring adequate oxygen and glucose supply to the brain and hence the brain integrity. Caregivers managing patients with cerebral pathology in the acute care setting should have adequate knowledge and thorough understanding of cerebral physiology and its alteration by certain drugs.

The main aim in management is to ensure that oxygen supply matches demand. Secondary brain damage resulting from ischemia is then minimized and ultimately neuronal function can be preserved. Pharmacotherapy can be used to decrease the metabolism and oxygen demands of the brain. Drugs can also be used to maintain cerebral blood flow through the control of cerebral perfusion pressure (CPP) and the caliber of cerebral vessels. In situations of brain injury and hypoxia, brain mass increases due to edema, increasing the intracranial pressure (ICP). This can be managed pharmacologically with the use of diuretics. Currently, drugs altering the mechanics of CSF production and absorption are not available for clinical use.

Drugs Modulating Cerebral Metabolism

In health, cerebral blood flow (CBF) is closely coupled to the metabolic activity of the brain. This is expressed as the cerebral metabolic rate for oxygen ($CMRO_2$) and is constant at 3.5 ml/100 g/min. This is to allow sufficient delivery of oxygen and glucose to the brain in order to meet the energy demands of the brain.

In brain injured patients, pharmacological agents that can reduce the $CMRO_2$ are often utilized so that the oxygen/energy needs can still be met in the face of decreased oxygen delivery because of decreased cerebral blood flow (Fig. 13.1).

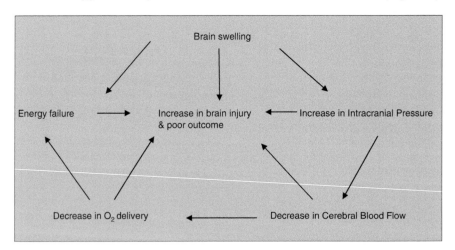

Fig. 13.1 Escalating cycle of brain swelling resulting in energy failure, increase in brain damage and poor outcome following injury to the brain (With permission of Wolters Kluwer Health; Decompressive craniectomy in head injury. Curr Opin Crit Care 2004: 10(2):101–4)

These agents mainly involve anesthetic drugs such as propofol, barbiturates and benzodiazepines. Volatile anesthetics reduce $CMRO_2$ while also exerting a dose-dependent vasodilatory effect on cerebral vessels that increases CBF.

Conversely, seizures and hyperthermia will increase $CMRO_2$ and subsequently CBF and ICP. If not treated appropriately, both conditions are associated with poorer outcome after successful cardiopulmonary resuscitation. Antipyretics such as acetaminophen can be used to manage hyperthermia, as well as physical methods of cooling. Seizures can be controlled with antiepileptics like phenytoin, benzodiazepine and barbiturates.

Propofol

This is a commonly used drug in anesthesia and critical care, mainly as an induction agent and in total intravenous anesthetic technique in neuroanesthesia. It is also used as a sedative agent in critically ill patients especially in neurocritical care unit to provide cerebral protection. Propofol decreases the cerebral metabolic rate for oxygen ($CMRO_2$), which leads to a reduction in CBF and ICP. However, one must be cautious when large doses are being used as it may result in systemic hypotension that will further compromise the CPP. Propofol is known to cause myoclonus but is non-epileptogenic, and is sometimes used to treat refractory status epilepticus.

Thiopental

Thiopental is less commonly used nowadays due to its unfavorable pharmacokinetic profile. It decreases $CMRO_2$ and causes cerebral vasoconstriction which result in reduced CBF, cerebral blood volume (CBV) and ICP. Thiopental can cause significant hypotension, which can impair CPP. It is sometimes used in selected cases with refractory elevated ICP to produce a state of deep coma when other conventional therapies failed to reduce ICP. Induced coma with the use of barbiturates like thiopental is called barbiturate coma.

Benzodiazepines

Agents in this category such as midazolam are widely used as sedative agents in neurocritical care unit. Its effect on cerebral physiology is similar to the above drugs, but to a lesser degree, with a less significant drop in $CMRO_2$, CBF, CBV and ICP.

Drugs Modulating Cerebral Blood Flow

Cerebral blood flow (CBF) is affected mainly by oxygen and carbon dioxide tensions in the blood and by autoregulation. CBF under normal conditions is maintained at a relatively constant rate despite variations in CPP. The mechanism involves alteration in the cerebral vascular resistance by changing the diameter of the cerebral arterioles. This phenomenon is known as autoregulation (Fig. 13.2). CBF is held constant over the CPP range of 50–150 mmHg. At the upper limit, the arterioles are maximally vasoconstricted and any increase in CPP will result in increased CBF. At the lower limit, the arterioles are maximally dilated in an attempt to maintain CBF, below which CBF becomes pressure dependent.

Cerebral perfusion pressure (CPP) depends on the mean arterial pressure (MAP) and the ICP or jugular venous pressure (JVP), whichever is greater, whereby

$$CPP = MAP - ICP \quad \text{or} \quad CPP = MAP - JVP$$

Many conditions can affect this autoregulatory process, which includes traumatic brain injury, hemorrhage, infection, tumor and certain pharmacological agents. Cerebral blood flow can be controlled by manipulating the mean arterial pressure and by changing the caliber of the cerebral blood vessels.

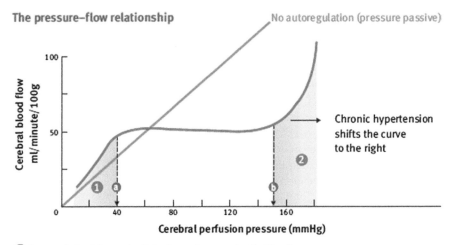

❶ Autoregulation fails: cerebral blood vessels are maximally dilated
❷ Autoregulation fails: cerebral blood vessels are maximally constricted

Between points ❸ and ❺ cerebral blood flow is kept relatively constant owing to changes in the diameter of the cerebral arterioles

Fig. 13.2 Autoregulation of blood flow in the brain depending on cerebral perfusion pressure (With permission from Surgery (Oxford); Applied physiology of the CNS Surgery 2005; 23 (1):7–12)

Drugs Affecting Mean Arterial Pressure

From the equation above, agents that increase mean arterial pressure will also increase cerebral perfusion pressure, which may help to maintain cerebral blood flow. So, drugs which are used to support the cardiovascular system by increasing the mean arterial pressure (see Chap. 11) are often used to try to improve cerebral blood flow in conditions of raised ICP. In the adult brain, the CPP is normally between 70 and 90 mmHg.

Drugs Affecting Caliber of Cerebral Blood Vessels

Drugs which alter the radius of cerebral blood vessels are not commonly used to improve cerebral blood flow and oxygenation in cases of brain injury. The following drugs are used in the management of brain injury in specific situations.

Nimodipine

Nimodipine, a cerebro-selective calcium channel blocker is administered prophylactically to prevent the development of cerebral vasospasm in patients who present with subarachnoid hemorrhage (SAH) due to ruptured aneurysm. Its use has been associated with improved outcome but mortality remains unchanged. It reduces the influx of calcium through the L-type calcium channels in smooth-muscle cells leading to decreased smooth-muscle contraction resulting in reduced arterial narrowing. In more recent studies, nimodipine has been applied intra-arterially for the treatment of refractory cerebral vasospasm.

Volatile or Inhalational Anesthetics

Volatile anesthetic agents are generally used only in the anesthetic management of brain injury, but will be mentioned here because of their significant effects on the brain. These agents are known to cause a dose-dependent rise in CBF as a result of cerebral vasodilation, which can lead to increase in CBV and ICP. Commonly used anesthetic agents such as isoflurane, sevoflurane and desflurane also cause a reduction in the $CMRO_2$ which leads to an uncoupling of flow and metabolism which is also concentration-dependent. Cerebral vascular reactivity to carbon dioxide and autoregulation are generally preserved but are gradually abolished by increasing concentration of volatile anesthetic agents. Enflurane at high concentrations and in the presence of hypocarbia causes epileptiform activity on electroencephalography, thus is not commonly used in neuroanesthesia. Desflurane when used for a prolonged period has been shown to increase ICP possibly due to increases in cerebrospinal fluid production. The use of nitrous oxide (N_2O) in neuroanesthesia is becoming rare due to its potent vasodilatory effect resulting in an increase in CBF. It also increases $CMRO_2$, therefore causing a clinically significant rise in ICP. Nitrous oxide should be avoided in patients with severely reduced intracranial compliance, in patients at risk of developing venous air embolism and pneumocephalus, during ventricular drainage and in the presence of air in the ventricles.

Other Drugs Affecting the Brain

Dexmedetomidine

Dexmedetomidine, a highly selective centrally acting α_2-adrenoceptor agonist is increasingly used as a sedative agent in intensive care unit (ICU) and is currently approved by Food and Drug Administration for short-term sedation of less than 24 h, though many studies have reported its safe use for longer duration. It has neuroprotective effects and is considered the near-ideal sedative agent in the ICU, due to its ability to maintain a patient's cooperation with minimal effect on respiratory function. Many studies had been carried out to evaluate its effect on CBF, $CMRO_2$ and ICP. Dexmedetomidine was found to have no effect on ICP. It had been shown to cause a dose-related reduction in both CBF and $CMRO_2$ in healthy subjects. However, these study results do not confirm that dexmedetomidine will not cause adverse effects on the CBF to CMR ratio in patients with cerebral pathology. Large randomized trials are still required to assess its safety in patients with neurologic diseases.

Magnesium Sulfate

This agent, commonly used in the management of eclampsia in obstetrics, is increasingly being used in neuroanesthesia due to its neuronal membrane stabilizing effect that may prevent seizure occurrence. There is evidence to show that the outcome is improved when it is used in patients with SAH. The postulated mechanism is that it lowers the intracellular smooth muscle calcium levels causing cerebral vasodilation.

Opioids

Opioids are commonly used in combination with sedatives or hypnotic agents to provide analgesia in patients undergoing neurosurgery and in patients who require invasive mechanical ventilation in the neurocritical care unit. They have minimal effects on CBF and $CMRO_2$, and carbon dioxide reactivity is preserved at standard clinical doses. However, if hypoventilation occurs and $PaCO_2$ is raised secondary to opioid use, cerebral vasodilation ensues which later results in increased CBF, CBV and ICP.

Drugs Used to Treat Cerebral Edema

Mannitol

This is an osmotic diuretic widely used to treat patients with cerebral edema irrespective of its etiology. Its use is supported by the Brain Trauma Foundation for the management of raised ICP after traumatic brain injury (TBI). There are two mechanisms involved. The first one occurs immediately after a bolus administration through plasma expansion, which decreases blood viscosity resulting in improved cerebral microvascular flow and consequently oxygen delivery to the brain. The second mechanism is due to its osmotic effect in which an osmotic gradient is established between the brain cells and plasma, causing water to be drawn out from the cells thereby reducing cerebral edema. Other than these two mechanisms, mannitol also acts as a free-radical scavenger. It decreases the harmful effect of free radicals during ischemia-reperfusion injury.

Hypertonic Saline

Hypertonic saline is an alternative to mannitol. There is some evidence that it has a more favorable effect compared to mannitol on mortality after TBI but larger trials are needed to prove this effect. Hypertonic saline encourages water to move out of brain cells into intravascular space, reducing cerebral swelling and ICP. Three percent saline can be given as continuous infusion or higher concentration solution is administered as bolus. The patient's volume status and plasma sodium concentration should be closely monitored.

Drugs Affecting CSF Production and Absorption

While these drugs do have an effect on CSF production and absorption, they are not used clinically in the management of brain injury. Certain drugs, particularly the volatile anesthetics, alter the mechanics of CSF production and absorption. For instance, isoflurane decreases the production and increases reabsorption of CSF. Sevoflurane causes a decreased in production but does not affect reabsorption. Prolonged administration of desflurane has been shown to cause significant rise in ICP, which is likely to be caused by increased CSF production. All opioids at low doses produce small increase in CSF absorption.

If the production of CSF is increased without any change in absorption, the increased volume of CSF within the skull will result in a rise in ICP. There are agents that are still in the experimental stage that can affect CSF outflow mechanics and potentially be utilized to reduce raised intracranial pressure.

Key Concepts

- Knowledge of cerebral physiology and its alteration by drugs is vital for care-givers managing patients with cerebral pathology.
- In patients with raised ICP, CBF needs to be maintained to meet metabolic demand and prevent secondary brain damage.
- Cerebral pathology can be managed by decreasing $CMRO_2$, maintaining adequate cerebral blood flow and oxygenation, and decreasing ICP.

Summary

Integrity of the brain is essential for meaningful survival. Therefore, it is crucial for all caregivers to understand the physiological process in the brain in a healthy person and its changes in disease states in order to provide appropriate and timely care. Early pharmacological modulation of abnormal physiological parameters is vital to ensure adequate blood flow to the brain so that oxygen and energy needs are met. While pharmacological measures are important in the management of brain pathology, other therapy including ventilation and oxygenation of the brain injured patient, and control of ICP via drainage of CSF play a big role. The ultimate aim is to prevent neuronal dysfunction and irreversible brain damage.

Further Reading

1. Brain Trauma Foundation, American Association of Neurological Surgeons, Congress of Neurological Surgeons. Guidelines for the management of severe traumatic brain injury. J Neurotrauma 2007;24 Suppl 1:S1–106.
2. Brunton L, Blumenthal D, Buxton I, Parker K, editors. Goodman & Gilman manual of pharmacology and therapeutics. 11th ed. New York: McGraw-Hill; 2008.
3. Fitch W. Physiology of the cerebral circulation. Best Pract Res Clin Anaesthesiol. 1999;13(4):487–98.
4. Glasby MA, Myles LM. Applied physiology of the CNS. Surgery. 2005;23(1):7–12.
5. Hutchinson PJ, Kirkpatrick PJ. Decompressive craniectomy in head injury. Curr Opin Crit Care. 2004;10:101–4.
6. Jantzen JP. Prevention and treatment of intracranial hypertension. Best Pract Res Clin Anaesthesiol. 2007;21(4):517–38.
7. Rang HP, Dale MM, Ritter JM, Flower RJ, Henderson G. Rang and Dale's pharmacology. 7th ed. Edinburgh: Churchill Livingstone; 2012.

Chapter 14
Drugs and the Liver/Gastrointestinal System

Li Lian Foo

Abstract Liver dysfunction is common in acutely ill patients. Unlike renal dysfunction where glomerular filtration rate (GFR) can be used to guide dose adjustments, there is no corresponding parameter to gauge the degree of liver dysfunction. Drug dosages have to be adjusted based on knowledge of pharmacodynamics and pharmacokinetics in liver disease. Metabolism of drugs with high hepatic extraction ratio (HER) is mainly dependent on blood flow to the liver. With cirrhosis and portosystemic shunting, blood flow to the liver is reduced and so is metabolism of high HER drugs. Oral bioavailability is increased and dosages should be reduced to prevent accumulation. Metabolism of low HER drugs is mainly dependent on protein binding and the intrinsic metabolic capacity of the liver. Drugs which are highly protein bound have significantly higher unbound concentrations in liver dysfunction. This is due to several mechanisms – reduced synthesis of albumin and alpha-1 acid glycoprotein, accumulation of endogenous compounds such as bilirubin which interfere in plasma protein binding, and possible qualitative changes in albumin and alpha-1 acid glycoprotein. Unbound drugs are biologically active, leading to more rapid onset of action and also increased adverse effects. For water soluble drugs, ascites and edema increases the volume of distribution and larger doses may be needed. The gastrointestinal tract is subject to many changes in the acutely ill patient. Gastrointestinal motility may be reduced due to pain, hypotension, hypoxemia, sepsis or the use of drugs such as dopamine and opioids. Hypoalbuminemia in the ill patient leads to mucosal edema and impaired absorption of enteral feeds and drugs. Acutely ill patients are also at increased risk of developing stress ulcers, which may lead to clinically important bleeding. This may be prevented with the use of appropriate stress ulcer prophylactic drugs.

Keywords Pharmacokinetics • Pharmacodynamics in liver dysfunction • Dose adjustments • Hepatic extraction ratio • Opioid dose • Stress ulcers in critically ill

L.L. Foo, M.D., M.Anaes (✉)
Department of Anesthesiology, Faculty of Medicine, University of Malaya,
50603 Kuala Lumpur, Malaysia
e-mail: foolilian@ummc.edu.my

© Springer International Publishing Switzerland 2015   119
Y.K. Chan et al. (eds.), *Pharmacological Basis of Acute Care*,
DOI 10.1007/978-3-319-10386-0_14

Introduction

Liver dysfunction is common in acutely ill patients. Eleven to fifty percent of critically ill patients in the ICU have liver dysfunction. This can be due to metabolic, inflammatory and hemodynamic factors such as sepsis, hypovolemic shock, or multi-organ dysfunction. With liver impairment, drug pharmacokinetics and pharmacodynamics are altered. It is important for the clinician to recognise liver dysfunction and choose appropriate drugs. Unlike renal dysfunction where GFR can be used to guide dose adjustments, there is no corresponding parameter to gauge the degree of liver dysfunction. Dosages have to be adjusted based on a sound knowledge of pharmacodynamics and pharmacokinetics in liver disease.

Similarly the gastrointestinal tract in the acutely ill patient undergoes paresis due to blood flow being redistributed away from the tract. Oral ingestion and absorption of drugs through this tract is therefore not reliable especially during episodes of vomiting and diarrhea. In addition with poor peristalsis, gastric contents accumulate and these may predispose to collections of fluid which can be aspirated into the lungs.

Liver Dysfunction

Absorption and Distribution

Drugs given enterally undergo first pass metabolism in the liver. High hepatic extraction ratio drugs (HER > 0.7) are taken up extensively and the amount of drug metabolised is mainly dependent on blood flow to the liver. On the other hand, metabolism of low hepatic extraction ratio drugs (HER < 0.3) is mainly dependent on the intrinsic metabolic capacity of the liver and protein binding (Table 14.1).

In liver cirrhosis portosystemic shunting reduces blood flow to the liver. High HER drugs will be less metabolised and have higher bioavailability. Therefore doses should be reduced. For example, oral bioavailability of carvedilol in cirrhosis is increased by 4.4 fold and midazolam by two fold.

In liver dysfunction, drugs which are highly protein bound have significantly higher unbound concentrations. This is due to several mechanisms – reduced synthesis of albumin and alpha-1 acid glycoprotein, accumulation of endogenous compounds such as bilirubin which interfere in plasma protein binding, and possible qualitative changes in albumin and alpha-1 acid glycoprotein. Unbound drugs are biologically active, leading to more rapid onset of action and also increased adverse effects. Highly protein bound drugs such as aspirin, benzodiazepines and phenytoin are most affected and dosage should be reduced. For water soluble drugs, ascites and edema increases the volume of distribution and larger doses may be needed.

Table 14.1 Examples of high and low hepatic extraction ratio drugs

Hepatic extraction	Low (<0.3)	Intermediate	High (>0.7)
	Warfarin	Quinidine	Cocaine
	Diazepam	Codeine	Morphine
	Phenytoin	Ondansetron	Lignocaine
	Theophylline	Nifedipine	Propanolol
	Carbamazepine	Aspirin	Verapamil

Metabolism and Excretion

Intrinsic hepatic clearance of drugs depends on metabolic enzyme capacity and activity of sinusoidal and canalicular transporters. In chronic liver disease, metabolism of drugs is decreased due to reduction in absolute cell mass, decrease in enzyme activity and impaired uptake of drugs across the endothelium. Drug doses should be reduced and interval between doses prolonged. In addition to impairment of biliary excretion, renal excretion of drugs may also be affected as advanced liver disease is often associated with hepatorenal syndrome and impaired renal function.

Acetaminophen and the Liver

Acetaminophen overdose is one of the commonest causes of acute liver failure. It is metabolized mainly in the liver into acetaminophen glucuronide (55 %) and acetaminophen sulfate (33 %). Five to ten percent is metabolized by hepatic P450 enzymes to N-acetyl-p-benzoquinoneimine (NAPQI), which is a highly reactive metabolite. Under normal circumstances NAPQI is detoxified by conjugation with glutathione and then excreted in urine.

In acetaminophen overdose, hepatic stores of glutathione may be depleted. Accumulation of NAPQI then leads to liver cellular necrosis. Factors increasing the risk of hepatotoxicity are poor nutritional status, increased age, certain genetic variations, chronic alcohol use, drugs which inhibit glucuronidation and drugs which induce liver P450 enzymes.

Patients are usually asymptomatic for the first 24 h. Subsequently they present with abdominal pain, nausea, and signs of liver failure. Treatment is by administration of N-acetylcysteine, which serves as a glutathione precursor, ideally within the first 16 h. N-acetylcysteine neutralizes NAPQI and prevents it from reacting with liver cells. Activated charcoal can be given to reduce acetaminophen absorption from the stomach if patients present within one hour of acetaminophen ingestion.

Opioids in Liver Disease

Opioids are used in the acutely ill for analgesia and sedation. The liver is the main site of metabolism for most opioids with the exception of remifentanil, which is cleared by ester hydrolysis. Opioids that are metabolized by glucuronidation (e.g., morphine, oxymorphone, buprenorphine) are mostly highly extracted and have increased oral bioavailability in liver dysfunction, leading to higher drug levels. Most other opioids are metabolized by oxidation and their clearance is reduced to a variable extent. Scoring systems for severity of liver disease, such as Child-Pugh Score and Model for End-Stage Liver Disease (MELD), lack sensitivity and correlation with liver drug metabolizing capacity. In severe liver failure, the concentration of opioids in the central nervous system may be increased, due to increased cerebral blood flow, increased permeability of the blood brain barrier and altered transporter function. This increases the risk of adverse effects such as respiratory depression and hepatic encephalopathy. Dosages of opioids have to be reduced, dosing intervals prolonged and sustained-release forms avoided in severe liver disease to reduce the risk of drug accumulation.

Gastrointestinal Dysfunction

The gastrointestinal tract is subject to many changes in the acutely ill patient. Gastrointestinal motility may be reduced due to pain, hypotension, hypoxemia, sepsis or the use of drugs such as dopamine and opioids. Hypermotility may be accompanied by diarrhea or vomiting. Hypoalbuminemia in the ill patient leads to mucosal edema and impaired absorption of enteral feeds and drugs. Acutely ill patients are also at increased risk of developing stress ulcers. Seventy five to one hundred percent of critically ill patients demonstrate evidence of stress-related mucosal disease within 24 h of admission to the intensive care unit and this may lead to clinically important bleeding.

Stress Ulcer Prophylaxis

Histamine 2-Receptor Antagonists

These drugs such as ranitidine and famotidine act by competitive inhibition of the histamine H_2 receptor on the gastric parietal cell, therefore reducing basal and meal-stimulated acid secretion. They are only moderately effective in the prevention of true stress ulcers. There are concerns regarding development of tolerance, leading to reduction in antisecretory effect within as early as 42 h despite the use of high dose regimens.

Proton Pump Inhibitors

Omeprazole and pantoprazole are the most commonly used drugs in this category. They are potent antisecretory drugs and provide more rapid and sustained increase in gastric pH than the histamine H_2 receptor antagonists. They act by forming a covalent disulfide link with a cysteinyl residue in the H^+/K^+-ATPase pump on the luminal surface of the gastric parietal cell. This results in irreversible blocking of the proton pump, and reducing the acid secretion by more than 90 %. Tachyphylaxis has not been reported. Individual inhibitors have different potential for drug interactions. Omeprazole may reduce the clearance of carbamazepine, diazepam and phenytoin. Pantoprazole has not shown clinically significant drug interactions and can be used without dose adjustment in patients with renal failure or moderate hepatic impairment.

Other Stress Ulcer Prophylactic Agents

Other therapies include *antacids* such as sodium citrate, *prostaglandin analogs* such as misoprostol, and sucralfate, which forms a protective barrier coating the gastric mucosa.

Key Concepts

- Liver dysfunction is common in the acutely ill patient.
- Adjustment of dose is needed in liver dysfunction based on the degree of liver impairment, protein binding and hepatic extraction ratio.
- Gastrointestinal dysfunction is similarly common in the acutely ill patient.
- Gastrointestinal bleeding carries significant morbidity and should be prevented with the use of appropriate stress ulcer prophylactic drugs.

Summary

Drug pharmacokinetics and pharmacodynamics are altered to a variable extent in liver disease. Unlike renal impairment, no single test exists to determine the degree of reduction in liver metabolic capacity. Dose adjustments have to be made based on the knowledge of the protein binding of each drug, the hepatic extraction ratio and the degree of liver impairment. The acute care physician should be especially cautious when using drugs with a narrow therapeutic index and those which may precipitate hepatic encephalopathy. The gastrointestinal tract is subject to many changes in the acutely ill patient. An increased susceptibility to stress ulcers and bleeding may increase mortality and should be prevented with stress ulcer prophylaxis.

Further Reading

1. Davis M. Cholestasis and endogenous opioids-liver disease and endogenous opioids pharma-cokinetics. Clin Pharmacokinet. 2007;46:825–50.
2. Fennerty MB. Pathophysiology of the upper gastrointestinal tract in the critically ill patient: rationale for the therapeutic benefits of acid suppression. Crit Care Med. 2002;30(Suppl): S351–5.
3. Smith L. Drug dosing considerations for the critically ill patient with liver disease. Crit Care Nurs Clin North Am. 2010;22:335e40.
4. Soultati A, Dourakis SP. Liver dysfunction in the intensive care unit. Ann Gastroenterol. 2005;18(1):35.
5. Twycross R, Pace V, Mihalyo M, Wilcock A. Acetaminophen (Paracetamol). J Pain Symptom Manage. 2013;46:747–55.
6. Verbeeck RK. Pharmacokinetics and dosage adjustment in patients with hepatic dysfunction. Eur J Clin Pharmacol. 2008;64(12):1147–61.

Chapter 15
Drugs and the Renal System

Sukcharanjit Singh Bakshi Singh and Jeyaganesh Veerakumaran

Abstract Renal impairment can be classified as acute or chronic renal disease. Renal impairment has significant effects on drug pharmacokinetics and pharmacodynamics. All aspects of drug pharmacokinetics can be affected (absorption, distribution, metabolism and excretion). Drug absorption will be reduced due to delayed gastric emptying and also due to gut edema. Drug distribution is altered due to changes in extracellular fluid volume, plasma protein binding and tissue binding of drugs. There is a decrease in drug metabolism in renal disease due to decreased activity of liver enzymes or alteration in activities of uptake and efflux transporters. Drug excretion in renal disease is affected due to changes in glomerular filtration, tubular secretion or reabsorption. Drug therapy in renal impairment should be individualized according to the extent of renal impairment and tendency of a drug used to cause nephrotoxicity. The route of elimination from the body and therapeutic index of the drugs are also important considerations.

Keywords Acute renal disease • Chronic renal disease • Renal impairment • Nephrotoxicity • Therapeutic index

Introduction

Renal impairment has significant effects on drug pharmacokinetics and pharmacodynamics. All aspects of drug pharmacokinetics, i.e., absorption, distribution, metabolism and excretion can be affected. This will result in blood and tissue drug concentrations that are different from what is expected in normal healthy individuals.

There are pharmacological studies done on renal impaired patients and these show that drug doses have to be adjusted in line with the degree of renal impairment. It is also important to know the degree by which these drugs are eliminated in the kidneys, in order to be able to make appropriate dose adjustments to achieve drug concentrations in the required therapeutic range, whilst avoiding toxic levels.

S.S. Bakshi Singh, M.D., M.Anaes (✉) • J. Veerakumaran, M.B.B.S., MMed Anaes
Department of Anesthesiology, Faculty of Medicine, University of Malaya, 50603 Kuala Lumpur, Malaysia
e-mail: sukcharan07@um.edu.my

© Springer International Publishing Switzerland 2015
Y.K. Chan et al. (eds.), *Pharmacological Basis of Acute Care*,
DOI 10.1007/978-3-319-10386-0_15

Classification of Renal Impairment

Renal impairment can be classified as acute or chronic renal disease. The degree of renal impairment can be calculated by measuring or estimating glomerular filtration rate (GFR). The stages of chronic kidney disease can be classified according to the National Kidney Foundation Kidney Disease Outcomes Quality Initiative (NKF KDOQI) guidelines (Table 15.1).

Creatinine clearance or glomerular filtration rate (GFR) can be estimated with the Cockroft–Gault equation, where:

$$\text{GFR} = \frac{(140 - \text{age}) \times \text{weight}}{72 \times \text{serum creatinine}} (\times 0.85 \text{ if the patient is female})$$

Acute renal failure has also been staged according to the mnemonic RIFLE, proposed by Acute Dialysis Quality Initiative (ADQI) group, as in Fig. 15.1.

Pharmacokinetics

Renal impairment can affect drug handling by the body in more than one way. It can affect drug absorption, drug distribution, metabolism and excretion.

Absorption

Drug absorption can be reduced due to delayed gastric emptying typically observed in patients with renal disease, especially end stage renal disease. Gastrointestinal symptoms such as vomiting and diarrhea can also affect drug absorption. Due to fluid retention in renal failure, edema of the gastrointestinal tract can also affect drug absorption. Patients with chronic renal failure might be on oral alkalizing agent (sodium bicarbonate, citrate) or antacids and these can cause decreased absorption of oral drugs that require an acidic environment (see Chap. 3).

Table 15.1 Stages of chronic kidney disease

Stage	Description	GFR (ml/min/1.73m^2)
1	Kidney damage with normal or ↑ GFR	≥ 90
2	Kidney damage with mild ↓ GFR	60 – 89
3	Moderate ↓ GFR	30 – 59
4	Severe ↓ GFR	15 – 29
5	Kidney failure	< 15 (or on dialysis)

Chronic kidney disease is defined as either kidney damage or GFR < 60 ml/min/1.73m^2 for 3 months. Kidney damage is defined as pathologic abnormalities or markers of damage, including abnormalities in blood, urine tests or imaging results (Permission from Elsevier; Am J Kidney Dis. 2002;39 2 Suppl 1:S1–246)

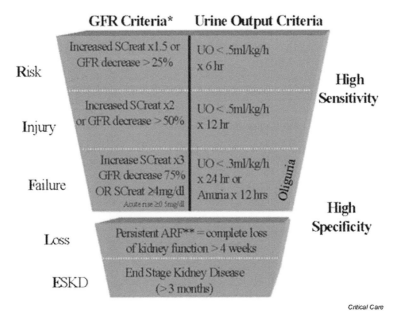

Fig. 15.1 Stages of acute renal failure according to the RIFLE classification. *ESKD* end stage kidney disease, *UO* urinary output, *ARF* acute renal failure (Taken from Crit Care. 2004;8:R204–12 (DOI 10.1186/cc2872))

Distribution

Renal impairment can affect drug distribution due to changes in extracellular fluid (ECF) volume, plasma protein binding and also tissue binding of drugs. Water soluble drugs are distributed in the ECF volume, hence an increase in the ECF associated with renal failure might result in requirement for an increased loading dose of water soluble drugs to produce the desired effect (see Chap. 4). Renal impairment can lead to metabolic acidosis and this can also alter tissue distribution of drugs.

The effect of a drug is attributed to the free fraction of the drug which is unbound from plasma protein. In renal disease, acidic drugs have decreased protein binding (usually to albumin) whereas basic drugs are usually unchanged or decreased. This effect is particularly important for drugs that are highly protein bound (>80%). The decrease in albumin or plasma protein binding can be attributed to the following causes:

(a) In the presence of uremia, plasma protein binding of acidic drugs is decreased. Most of the acidic drugs bind to albumin. It has been postulated that in the presence of renal disease with uremia, there might be displacement of drugs from albumin binding sites by organic molecules that accumulate in renal failure. The free fraction of highly protein bound drugs will increase and this may lead to drug toxicity.

(b) Renal failure can lead to albumin loss (nephrotic state), decreasing the amount of albumin available to bind to drugs which will lead to an increase in free fraction of the drugs. This is especially important in highly protein bound drugs such as phenytoin.
(c) Renal disease can also lead to a hypermetabolic state which can lead to increase protein breakdown (especially in stress, trauma or sepsis).
(d) Renal impairment can also cause structural changes in albumin preventing acidic drug from binding to it.

Hence, the decrease in protein binding can lead to increases in free fraction of drugs, resulting in a potential to cause drug toxicity (see Chaps. 4 and 10).

Metabolism

Drug metabolism in renal disease can decrease or remain unchanged. There is significant evidence that chronic kidney disease may lead to alteration in non-renal clearance of many medications as a result of alterations in activities of uptake and efflux transporters. In the presence of renal failure there is a down regulation of selected isoforms of hepatic cytochrome P450 (CYP) enzymes due to a decrease in gene expression. This is accompanied by major reduction in metabolism mediated by P450 enzymes (see Chap. 5). Intestinal metabolism of certain drugs is also affected in renal failure, as evidenced by increased bioavailability of several drugs, reflecting a decrease in either intestinal first-pass metabolism or extrusion of drugs (mediated by P-glycoprotein). Intestinal P450 is also down-regulated secondary to decreased gene expression. Both the processes of drug hydrolysis and reduction will be slower in patients with impaired renal function. This may be partially due to the kidney possessing the same metabolizing enzymes as the liver (example renal P450 enzyme). However, oxidation of drugs is not changed in chronic renal disease.

Excretion

Renal handling of drugs may involve glomerular filtration, tubular secretion or tubular reabsorption. Drugs with low plasma protein binding are filtered more readily. Large molecules (MW > 20,000 Da) are not readily filtered. Excretion of drugs in kidney disease is significantly reduced. This can affect both parent drug and drug metabolites (especially meperidine, morphine, procainamide) and can lead to toxicity.

Any factor that leads to a decrease in ECF volume can increase the risk of worsening renal impairment. These include decreased fluid intake, use of diuretics or loss of body fluids (vomiting, diarrhea, bleeding). Other factors such as decreased renal blood flow, decreased GFR and decreased secretion of drugs can also cause a decrease in drug excretion and will promote drug accumulation leading to toxicity (see Chap. 5).

Drug Therapy in Renal Impairment

Drug therapy should be individualized according to the extent of renal impairment. Creatinine clearance should be measured to determine GFR. Dependence on serum creatinine alone as a measure of renal function might be misleading as it is affected by other factors such as muscle mass of a patient. It is also not reliable in elderly patients due to decreased muscle mass and a decrease in GFR. Some drugs (cimetidine, trimethoprim) can cause an increase in serum creatinine.

Drug selection should be guided by renal function, the mode of excretion of the drug and also the effect of the drug on renal function. Many drugs (e.g., NSAIDs) can affect renal function and should be used with caution in high risk patients. Some drugs are excreted exclusively by the kidneys (e.g., aminoglycosides, lithium) so care must be taken to adjust the drug dosing protocol and monitor for signs of toxicity in the presence of renal dysfunction.

The therapeutic index (TI) of drugs should be known prior to the use of a drug. Therapeutic index of a drug is determined by dividing median toxic dose (TD_{50}) with median effective dose (ED_{50}). In drugs with narrow therapeutic index, a small change in drug concentration may result in toxicity or loss of efficacy.

Drugs can be safely used if the following guidelines are followed:

1. The use of a loading dose is probably best avoided in patients with renal failure, especially for drugs with narrow therapeutic index.
2. The maintenance dose of a drug can be reduced or the dosing interval be lengthened to reduce the risk of drug toxicity. The maintenance dose reduction method is used whenever a more constant serum drug level is therapeutically preferable (e.g., β-lactam antibiotics), whereas the interval extension method is used for drugs for which a constant serum level is either unnecessary (e.g., vigabatrin) or undesirable (e.g., aminoglycoside antibiotics).
3. Use TDM (Therapeutic Drug Monitoring) and monitor renal function when using nephrotoxic drugs.
4. Avoid nephrotoxic drugs where possible (e.g., aminoglycosides, amphotericin B, cisplatin)
5. Avoid dehydration.
6. Use smaller pre-calculated doses or longer intervals in patient with kidney disease.

Key Concepts

- The stages of acute or chronic renal failure are best classified and described in terms of an estimated glomerular filtration rate.
- Renal failure has an impact on the absorption, distribution, metabolism and more significantly excretion of most drugs.

- Therapeutic Drug Monitoring and monitoring of renal function, when using nephrotoxic drugs are important for safe administration.
- Avoid nephrotoxic drugs where possible (aminoglycosides, amphotericin B, cisplatin) in all renal impaired patients.
- Avoid dehydration in renal patients as this will worsen the renal status and the impact of the administered drugs.

Summary

There are significant changes in drug pharmacokinetics and pharmacodynamics in patients with renal impairment. All aspects of drug pharmacokinetics (absorption, distribution, metabolism and excretion) are affected in renal impairment. Drug therapy should be individualized and should take into account the extent of renal impairment, the effect of the drug on the kidneys and the route of elimination of the drug used.

Further Reading

1. Atkinson A. Effects of renal disease on pharmacokinetics. 1999. Available from http://faculty.ksu.edu.sa/hisham/Documents/Students/a_PHCL/renal.pdf. Last accessed 22 June 2014.
2. Bellomo R, Ronco C, Kellum JA, Mehta RL, Palevsky P, The Other ADQI Workgroup. Acute renal failure – definition, outcome measures, animal models, fluid therapy and information technology needs: the Second International Consensus Conference of the Acute Dialysis Quality Initiative (ADQI) Group. Crit Care. 2004;8:R204–12. doi:10.1186/cc 2872. Available from http://ccforum.com/content/8/4/R204
3. Bowman L, Luppa J. Principles of drug dosing in renal impairment. In: Cheng S, Vijayan A, editors. The Washington manual[TM]: Nephrology. 3rd ed. Alphen aan den Rijn: Publisher under Wolters Kluwer (Lippincott Williams and Wilkins). 2012. ISBN 9781451114256.
4. Cockcroft DW, Gault MH. Prediction of creatinine clearance from serum creatinine. Nephron. 1976;16:31–41.
5. National Kidney Foundation (NKF) Kidney Disease Outcome Quality Initiative (K/DOQ1) Advisory Board. K/DOQ1 clinical practice guidelines for chronic renal diseases: evaluation, classification and stratification. Kidney Disease Outcome Quality Initiative. Am J Kidney Dis. 2002;39(2 Suppl 1):S1–246.
6. Pichette V, Leblond FA. Drug metabolism in chronic renal failure. Curr Drugs Metab. 2003;4 (2):91–103. Available from www.ncbi.nlm.nih.gov/pubmed/12678690. Last accessed on 16 June 2014.
7. Reindenbery MM, Drayer DE. Alteration of drug protein binding in renal disease. Clin Pharmacokinet. 1984;9(Suppl):18–26.
8. Sun H, Frassetto L, Benet LZ. Effects of renal failure on drug transport and metabolism. Pharmacol Ther. 2006;109(1–2):1–11. Epub 8 Aug 2005. Available from www.ncbi.nlm.nlm.gov/pubmed/16085315) Last accessed on 16 June 2014
9. Verbeek RK, Musuamba FT. Pharmacokinetics and dosage adjustment in patients with renal dysfunction. Eur J Clin Pharmacol. 2009;65(8):757–73. doi:10.1007/s00228-009-0678, Epub 20 Jun 2009

Chapter 16
Drugs and the Endocrine System

Nur Lisa Zaharan and Pui Kuan Lee

Abstract The endocrine system regulates many aspects of important physiological processes in the body. Our body produces hormones from different glands and these hormones are kept in balance by various regulatory feedback mechanisms. Autoimmune problems, tumors, metabolic dysregulation and genetic predisposition may disrupt normal regulation of hormones causing excessive production, hormone deficiency or resistance of target organs to hormones. Three hormones, i.e., thyroid hormones, insulin and glucocorticoids are at issue more often compared to the rest of the hormones in the endocrine system. In hyperthyroidism, various antithyroid drugs are used, which target the different stages in thyroid hormone synthesis and release. In hypothyroidism, thyroxine is used in replacement therapy. Diabetes mellitus is an increasing problem worldwide where high blood glucose level is attributed to the deficiency of insulin (Type 1 diabetes) or insulin resistance (Type 2 diabetes.) Type 1 diabetes is managed with human insulin or insulin analogues. In Type 2 diabetes, where there is insulin resistance and relative insulin deficiency, insulin secretagogues and insulin sensitizers are used. Newer agents target the incretin hormones which play an important role in the regulation of insulin, glucose and satiety. The adrenal glands which produce the glucocorticoids and mineralo-corticoids regulate many metabolic processes and are important in inflammatory and immune responses. Synthetic glucocorticoids are used clinically in a wide variety of clinical conditions not related to hormone deficiencies due to its potent anti-inflammatory and immunosuppressant effects. However, the use of glucocorticoids therapeutically is associated with many adverse effects, among them impaired response to infection and Cushing's syndrome.

Keywords Hormone replacement • Antithyroid drug • Thyroid hormone replacement • Insulin therapy • Oral antidiabetic agents • Corticosteroids

N.L. Zaharan, M.B.B.Ch, B.A.O., B.Med.Sc., Ph.D. (✉)
Department of Pharmacology, Faculty of Medicine, University of Malaya,
50603 Kuala Lumpur, Malaysia
e-mail: lisa@ummc.edu.my

P.K. Lee, M.D., M.Anaes
Department of Anesthesiology, Faculty of Medicine, University of Malaya,
50603 Kuala Lumpur, Malaysia

© Springer International Publishing Switzerland 2015   131
Y.K. Chan et al. (eds.), *Pharmacological Basis of Acute Care*,
DOI 10.1007/978-3-319-10386-0_16

Introduction

The endocrine system is important in regulating metabolic processes in the body. The hormones may act on an organ or a specific metabolic process. Most of the endocrine organs are controlled by feedback mechanisms either through stimulating hormones produced centrally by the hypothalamus or pituitary, or through the effects of the index hormone. Dysfunction in endocrine system results in either a deficiency or an over-production of the hormone, causing significant morbidity. When the hormone levels are elevated, pharmacological strategies are required to bring down the levels and to treat the effects of these elevated hormones. On the other hand, if the hormone levels are low, replacement therapy or therapy to stimulate hormone production is instituted. Sometimes, the target organ becomes resistant to the effect of the hormone, in which case drugs are used to improve the sensitivity of target organs to the hormone. Synthetic hormones such as glucocorticoids are also used therapeutically in clinical conditions not related to hormone deficiency.

Thyroid Hormones

The thyroid gland synthesizes two hormones, triiodothyronine (T_3) and thyroxine (T_4). In the process of synthesizing thyroid hormones, it utilizes iodine. The release of T_3 and T_4 are controlled by hypothalamic thyrotropin-releasing hormone (TRH) and thyroid-stimulating hormone (TSH) from the pituitary, which are inhibited in the presence of T_3 and T_4. This is a negative feedback mechanism. More than 99 % of thyroid hormones are bound to proteins such as the thyroid-binding globulin (TBG). Free forms of T_4 and T_3 dissociate from TBGs and enter cells via diffusion or active transport. Within the cells, T_4 is converted to T_3, which is the active form. T_3 enters the nucleus, activating the transcription of genes, causing the synthesis of proteins and other substances. Thyroid hormones regulate carbohydrate, protein and lipid metabolism and are required for growth and development. The overall effect in cellular metabolism is to increase oxygen consumption and basal metabolic rate. In addition, thyroid hormones also have an effect on the sympathetic nervous system.

Drugs Used in Management of Hyperthyroidism

In hyperthyroidism, drugs are administered to stop the production of thyroid hormone, prevent its release from the thyroid gland and block the adrenergic effects of excessive hormonal levels.

Antithyroid Drugs (Thioureylenes)

Thioureylenes (also known as thioamides) include carbimazole and its active metabolite, methimazole, as well as propylthiouracil (PTU). These drugs block de novo synthesis of thyroid hormone by inhibiting the iodination process of tyrosine residues in thyroglobulin. Tyrosine residues are iodinated to form monoiodotyrosine (MIT) and in some, di-iodotyrosine (DIT). These molecules are coupled in pairs, MIT with DIT to form T_3, or two DIT molecules to form T_4. PTU has the extra advantage of inhibiting the peripheral conversion of T_4 to T_3. Stored T_3 and T_4 are unaffected. The clinical effects of carbimazole and PTU may be delayed until the thyroid hormone stores are depleted, which may take 2–4 weeks. The well-known but rare and potentially fatal adverse effect of thioureylenes is agranulocytosis, which renders the patients prone to infection. Agranulocytosis is reversible when the drug is stopped. PTU is associated with hepatotoxicity. Long-term or high-dose treatment with these agents may cause hypothyroidism. Thioureylenes cross the placenta and should be used with caution in pregnant patients.

Iodides

Potassium iodide (Lugol's iodine) is given orally and is converted to iodide in vivo. It inhibits the synthesis and release of thyroid hormone and reduces the vascularity of the thyroid gland. Iodides also block peripheral conversion of T_4 to T_3. Lugol's iodine is used in thyroid crisis and before thyroidectomy in vascular goiters.

Radioiodine ^{131}I

^{131}I emits beta and gamma radiation. The beta rays are absorbed by thyroid tissues and are cytotoxic. The cytotoxic effect is quite delayed with maximal effect observed around 4 months post-treatment. Hypothyroidism is commonly seen after treatment, especially in patients with Grave's disease. Sialadenitis which manifest as xerostomia, altered taste or pain is another adverse effect. Radioiodine should be avoided in children, pregnant and nursing women due to the radiation.

Beta Adrenoceptor Blockers

Beta blockers used in the management of hyperthyroidism include propranolol and esmolol. They block the peripheral adrenergic manifestations of hyperthyroidism such as palpitations, tremors and anxiety. Propranolol has additional function in hyperthyroidism as it can block peripheral conversion of T_4 to T_3.

Hyperthyroid Emergencies

Thyrotoxic crisis or thyroid storm is an extreme form of thyrotoxicosis with severe symptoms including tachycardia, vomiting, diarrhea, dehydration and delirium. Urgent treatment is required as mortality is high. Reduction of thyroid hormone is achieved by giving thioureylenes (PTU or carbimazole). Potassium iodide is given 1 h after thioureylenes (to prevent the iodine from being used as a substrate for new thyroid hormone synthesis) to produce a complete block of thyroid hormone production. Beta blockers are given to treat autonomic symptoms and hydrocortisone administered to treat hypoaldosteronism which is associated with Grave's disease.

Drugs Used in the Management of Hypothroidism

Hypothyroidism can be due to a defect in the hypothalamus, the pituitary gland or the thyroid gland. The standard thyroid replacement therapy is levothyroxine (T_4), which is available both in oral and intravenous form ($t_{1/2}$ 7 days) and liothyronine (T_3) which is available in the oral form ($t_{1/2}$ 2 days) and is four times more potent than levothyroxine. Liothyronine has a faster onset of action than levothyroxine.

Hypothyroid Emergencies

In myxedema coma, a hypothyroid emergency which is life threatening, clinical features of profound decrease in metabolic activity are seen, such as hypothermia, hypotension, bradycardia and depressed level of consciousness. Replacement thyroid hormone is given intravenously until the patient is able to take orally. Supportive therapy includes gradual rewarming and appropriate oxygen therapy as necessary to manage hypercapnia and hypoxia. Intravenous dextrose is given for hypoglycemia and hydrocortisone if adrenal insufficiency is present. Hypotension and hyponatremia may be corrected with saline infusion and associated infections must be treated.

Diabetes Mellitus

Type 1 diabetes mellitus (T1DM) is characterized by insulin deficiency while type 2 diabetes mellitus (T2DM) is commonly due to resistance of the principal cells to the effects of insulin. It can progress to insulin deficiency.

Insulin

Insulin is secreted by the beta cells of the pancreas. It is an anabolic hormone involved in the metabolism of carbohydrate, protein and lipid in the body and acts on three principal cells, i.e. muscles liver and adipose tissues. The main functions are to increase the storage of glucose (glycogenesis), reduce glucose production (gluconeogenesis) and decrease release of glucose (glycogenolysis) by the liver as well as to increase glucose uptake and utilization in adipose tissues. In addition, insulin increases lipid synthesis in the liver and adipose tissues whilst reducing lipolysis. Insulin is also important in protein breakdown in the liver and protein uptake and synthesis in the muscles.

Insulin Therapy for Diabetes Mellitus (T1DM and Some T2DM)

The application of genetic engineering enabled human insulin to be produced using recombinant DNA technology. In the 1990s, insulin analogues were developed and these analogues are now gradually replacing human insulin.

Insulin, being a protein, is destroyed in the gastrointestinal tract and so is administered parenterally, usually via subcutaneous injections. Once absorbed, insulin has a short half-life of 10 min. Insulin is given as replacement therapy in patients with T1DM and some patients with T2DM. Intravenous insulin is used as a short-term measure for situations like surgical procedures and insulin is also given to pregnant patients with diabetes mellitus.

Hypoglycemia is the most common adverse effect of insulin therapy. It is exacerbated by an inappropriately high dose of insulin, low carbohydrate intake and/or excess physical activity. Some patients may acquire 'hypoglycemic unawareness' in which autonomic signs of hypoglycemia are reduced or absent. Other adverse effects include weight gain, lipodystrophy at injections sites and insulin allergy (hypersensitivity). Insulin edema is a rare adverse reaction due to sodium and water retention seen at the start of treatment in patients with poor glucose control.

Insulin Preparations

There are four types of insulin preparations: rapid or ultrashort-acting, short-acting, intermediate-acting and long-acting as presented in Table 16.1. Short-acting insulin is given before meals to increase the level of circulating insulin and thus target postprandial hyperglycemia. Intermediate-acting insulin is produced by combining insulin with neutral protamine, also known as Neutral Protamine Hagedorn insulin

Table 16.1 Pharmacological properties of the different insulin preparations

Type of insulin	Examples	Onset	Peak activity	Duration
Rapid/Ultrashort-acting insulin	Insulin lispro, Insulin aspart	15–30 min	30–90 min	3–4 h
Short-acting insulin	Regular insulin	30–60 min	2–4 h	6–10 h
Intermediate-acting insulin	Neutral protamine Hagedorn (NPH insulin)	1–4 h	4–12 h	12–24 h
Long-acting insulin	Insulin glargine Insulin detemir Insulin glulisine	1–2 h	3–20 h	24–30 h

Permission from Hermansen, K. Insulin and new insulin analogues, insulin pumps and inhaled insulin in type 1 diabetes, in Pharmacotherapy of Diabetes: New development (eds: Mogensen, C) Springer 2007 New York

or NPH insulin. Intermediate-acting insulin is also available as combined or premixed preparation with fast-acting insulin. Long-acting insulin is used at bedtime or in the morning to provide basal insulin replacement.

Antidiabetic Agents for Type 2 Diabetes Mellitus

In patients with T2DM, the available drugs are used to stimulate insulin secretion using insulin secretagogues or to improve target organ sensitivity to insulin using insulin sensitizers. New drugs acting on incretins such as glucagon-like peptide-1 (GLP-1) have been developed. Incretins are gastrointestinal hormones that have glucose-lowering actions. GLP-1 stimulates glucose dependant insulin secretion and mediates the satiety response. Antidiabetic agents currently available are summarized in Table 16.2. Metformin is the first-line antidiabetic agent according to many guidelines due to its weight loss effects and evidence linking it to decrease in cardiovascular complications in diabetes.

Hyperglycemic Emergencies

In diabetic ketoacidosis, patients present with depressed mental status, dehydration, rapid deep breathing, and fluid, electrolyte and acid-base derangement. Management is based on the following four principles: (1) fluid and electrolyte therapy (2) intravenous insulin therapy (3) treatment of co-morbidities such as infections and (4) close monitoring. Electrolyte therapy includes management of potassium, in certain cases, bicarbonate and phosphate levels. IV Insulin should be adjusted regularly according to glucose levels.

Table 16.2 Antidiabetic agents for type 2 diabetes mellitus

Drug class	Drug name	Main mechanisms of action	Adverse effects/precaution
Biguanides	Metformin	Insulin sensitizer Increase glucose uptake by tissues Decrease hepatic gluconeogenesis Decrease glucose absorption from GIT	GI disturbance Lactic acidosis (Caution in those with moderate-severe renal impairment)
Sulfonylureas	Glibenclamide Gliclazide Glimepiride Glipizide	Insulin secretagogues Binds to sulfonylureas receptors associated beta cells K_{ATP} inward rectifier channels to stimulate production of insulin	Hypoglycemia Weight gain Many drug interactions Alcohol-flushing
Meglitinides	Repaglinide Nateglinide	Insulin secretagogues Binds to sulfonylureas receptors. MOA similar to sulfonylureas.	Hypoglycemia Drug interactions: Cytochrome P450
Thiazolidinedione	Pioglitazone	Insulin sensitizer Ligands of the peroxisome proliferator-activated receptor-gamma (PPAR-γ) receptors Increase glucose and lipid uptake and utilization by tissues	Fluid retention Weight gain Risk of osteoporosis Slight increased risk of bladder cancer
Incretin modulators	Exenatide*	Glucagon-like peptide-1 receptor agonists Action includes stimulating insulin release, suppress glucagon and decrease appetite	Reports of pancreatitis Hypoglycemia
	Sitagliptin	Dipeptidyl-peptidase-4-(DPP-4) inhibitor The DPP-4 enzyme is involved in inactivating GLP-1. Action is similar to GLP-1 agonist	Upper respiratory tract infections Drug interactions
α-glucosidase inhibitor	Acarbose	Reduces the absorption of carbohydrates in the intestine	GI disturbance

ᵃExenatide is administered via subcutaneous injections

Hypoglycemic Emergencies

Blood sugar level below 3.9 mmol/l (70 mg/dl) is defined as hypoglycemia. Clinical features include sweating and palpitations, and central nervous dysfunction such as blurred vision, headache and weakness which can progress to convulsions and coma. Patients with hypoglycemia can be given glucose orally or intravenously. In hypoglycemic emergencies outside hospital, when patients are unable to take orally, glucagon injections (packed in a kit) can be given, which will return blood glucose levels to normal in 10–15 min.

Adrenal Hormones

The adrenal glands produce glucocorticoids, mineralocorticoids and sex hormones. Glucocorticoids are important in the maintenance of the hypothalamic-pituitary-adrenal (HPA) axis. Glucocorticoids also play important roles in stress response and metabolic regulations. They possess potent anti-inflammatory and immunosuppressive properties. In acute inflammation, glucocorticoids decrease the activity of leucocytes while in chronic inflammation glucocorticoids reduce the clonal expansion and activity of lymphocytes. They also reduce angiogenesis and fibrosis associated with inflammation. In addition, glucocorticoids play an important role in regulating inflammatory mediators such as cytokines, prostaglandins, histamines and complement components.

The important mineralocorticoid synthesized by the adrenal glands is aldosterone. Aldosterone plays a vital role in salt and water balance in the body by regulating the renin-angiotensin-aldosterone system.

Synthetic Corticosteroids

Many synthetic corticosteroids have been developed with different potencies (Table 16.3). Corticosteroids possess potent anti-inflammatory and immunosuppressive properties and have become valuable treatment options for a wide range of clinical conditions. However, multiple adverse effects of corticosteroids have also been reported (Table 16.4). Most of the adverse effects of corticosteroids are related to the dosage given. In order to reduce adverse effects, the lowest possible dose and for the shortest duration of treatment that will produce the desired therapeutic response should be used. Steroid dose is tapered down gradually before being stopped. Sudden withdrawal of corticosteroids after prolonged or high dose may

Table 16.3 Potencies of different corticosteroids

Corticosteroids	Relative potency to hydrocortisone	Main route of administration	Duration of action
Hydrocortisone	1	IV	8–12 h
Prednisolone	4	Oral	12–36 h
Dexamethasone	30	IM, oral	36–54 h
Betamethasone	30	Topical	36–54 h

Table 16.4 Adverse effects of corticosteroids

Cushing's syndrome	Moon face, plethora, buffalo hump, purple striae, central obesity, poor wound healing, skin thinning, muscle wasting and weakness, avascular necrosis, increased appetite, insomnia, euphoria or dysphoria
Immunosuppression	Susceptibility to infections, poor wound healing, peptic ulcerations
Suppression of adrenal functions	Addisonian crisis
Metabolic	Hyperglycemia, hyperlipidemia, hypertension
Musculoskeletal	Osteoporosis, muscle wasting, proximal muscle weakness, disturbance of growth in children
Eye	Cataract, glaucoma
Central nervous system	Insomnia, euphoria, depression, steroid psychosis
Others	Hirsutism, dry mouth

precipitate acute adrenal insufficiency. Corticosteroids may be administered by many routes: oral, intravenous, intra-articular, inhalation, eye or nasal drops and topical skin application. They are transported in the plasma bound to corticosteroid-binding globulin (CBG) and to albumin. Some synthetic glucocorticoids are not bound to protein. They are metabolized by the microsomal enzymes in the liver.

Synthetic Mineralocorticoids

Fludrocortisone is a synthetic mineralocorticoid which is used as replacement therapy for adrenal insufficiency. It has a long duration of action (24–36 h) and is metabolized by the hepatic microsomal enzymes.

Adrenal Emergencies

In adrenal emergencies such as Addisonian's crisis, the adrenal insufficiency can cause headache, weakness, nausea and vomiting, diarrhea and confusion or coma. Patients present with hypotension, which when left untreated can progress to shock that does not respond to volume replacement and vasopressors, leading to death. Management involves immediate treatment with hydrocortisone 100–300 mg and aggressive volume replacement. Serum electrolytes and blood glucose should be monitored and corrected as necessary.

Key Concepts

- Excessive or deficiencies in hormones may present as acute and chronic medical problems.
- Replacement therapy is instituted for deficiencies of hormones.
- Other strategies in endocrine problems include targeting hormone synthesis and glands as well as targeting sensitivities of organs to hormones.
- Adverse effects of replacement therapy usually relates to its actions in the body.

Summary

Hormones control many physiological processes in our body and is kept in balance by various regulatory mechanisms. Conditions in which the hormone is deficient or target organ is resistant to hormone actions, as exemplified by hypothyroidism, diabetes mellitus and Addison's disease, require pharmacological replacement of the deficient hormone or pharmacological strategies to improve sensitivities to hormone. Conditions in which hormones became excessive and unregulated, as exemplified by thyroid storm, require pharmacological strategies to bring down the hormone levels by targeting various stages of hormone synthesis, release and actions. Hormones such as corticosteroids are also used for other clinical conditions for their potent anti-inflammatory and immunosuppressant effects as well as for diagnostic purposes.

Further Reading

1. Buchman A. Side effects of corticosteroid therapy. J Clin Gastroenterol. 2001;33(4):289–94.
2. Hampton J. Thyroid gland disorder emergencies: thyroid storm and myxedema coma. AACN Adv Crit Care. 2013;24(3):325–32.
3. Hermansen K. Insulin and new insulin analogues, insulin pumps and inhaled insulin in type 1 diabetes. In: Mogensen C, editor. Pharmacotherapy of diabetes: new developments improving life and prognosis for diabetic patients. New York: Springer; 2007.
4. Brent G, Koenig R (2011) Thyroid and anti-thyroid drugs. In: Brunton LL, Blumenthal DK, Murri N, Dandan RH, Knollmann BC (eds) Goodman & Gilman's the pharmacological basis of therapeutics, 12th edn. McGraw-Hill, New York
5. Inzucchi S, Bergenstal R, Buse J, Diamant M, Ferrannini E, Nauck M, et al. Management of hyperglycemia in type 2 diabetes: a patient-centered approach: position statement of the American Diabetes Association (ADA) and the European Association for the Study of Diabetes (EASD). Diabetes Care. 2012;35(6):1364–79.
6. Kearney T, Dang C. Diabetic and endocrine emergencies. Postgrad Med J. 2007;83 (976):79–86.
7. Peck TE, Hill SA, Williams M. Pharmacology for anaesthesia and intensive care. 2nd ed. London: Alden Group; 2003. Oxford.
8. Schäcke H, Döcke W, Asadullah K. Mechanisms involved in the side effects of glucocorticoids. Pharmacol Ther. 2002;96(1):23–43.
9. Stein S, Lamos E, Davis S. A review of the efficacy and safety of oral antidiabetic drugs. Expert Opin Drug Saf. 2013;12(2):153–75.
10. Torrey S. Recognition and management of adrenal emergencies. Emerg Med Clin North Am. 2005;23(3):687–702.
11. Van Ness-Otunnu R, Hack J. Hyperglycemic crisis. J Emerg Med. 2013;45(5):797–805.

Chapter 17
Drugs and the Neuromuscular System

Kang Kwong Wong

Abstract Acute neuromuscular disorders usually manifest as muscle weakness. The weakness can arise from primary neuromuscular disorders or can be a manifestation of a systemic illness. Few pharmacological agents act directly on the neuromuscular system. Most of the currently available agents work at the neuromuscular junction by inhibiting acetylcholinesterase. This prevents hydrolysis leading to increased concentrations of acetylcholine in the neuromuscular junction. Anticholinesterase agents are used in the management of patients in a limited number of neuromuscular disorders, for example in the management of myasthenia gravis and in the treatment of glaucoma. They are also used to reverse the effects of muscle relaxants.

Keywords Neuromuscular system • Neuromuscular junction • Neuromuscular disorders • Acetylcholine • Myasthenia gravis • Acetylcholinesterase

Introduction

The neuromuscular system consists of motor neurons, muscles and the neuromuscular junctions. There is no direct contact between the neurons and the muscles as they are separated by the neuromuscular junction. Important physiological processes occur in this junction in order for the neuromuscular system to work efficiently. Most drugs which act on the neuromuscular system exert their pharmacological actions through the neuromuscular junction.

Physiology of the Neuromuscular System

The motor neurons which control skeletal muscle are long cells of up to 1 m. The cell body is located in the ventral horn of the spinal cord. Information in the form of electrochemical signaling is transmitted down the axon. Axons are around

K.K. Wong, M.B.B.S., M.Anaes (✉)
Department of Anesthesiology, Faculty of Medicine, University of Malaya,
50603 Kuala Lumpur, Malaysia
e-mail: ricwkk@um.edu.my

© Springer International Publishing Switzerland 2015   141
Y.K. Chan et al. (eds.), *Pharmacological Basis of Acute Care*,
DOI 10.1007/978-3-319-10386-0_17

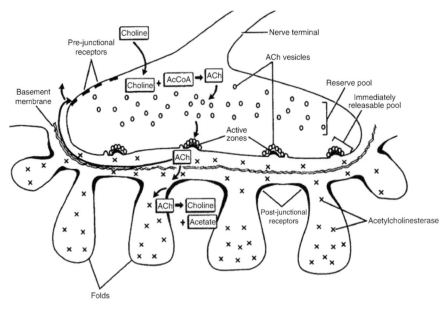

Fig. 17.1 Structure of neuromuscular junction (With permission from Oxford University Press; Br J Anaesth CEPD Rev. 2002;2(5):129–33)

10–20 μm in diameter and are surrounded by a myelin sheath. The myelin sheath is interrupted by gaps known as nodes of Ranvier. The nodes of Ranvier speed up propagation of action potential via a process known as saltatory conduction. Before reaching the neuromuscular junction, the axon of these nerves branches into several terminals to innervate many muscle cells.

Each muscle cell has only one neuromuscular junction (Fig. 17.1) and is only innervated by one nerve. A nerve and the muscle cells which it innervates together form one motor unit. The number of muscle cells per motor unit varies from a few to several thousand, depending on the function of the muscle. Strong bulky muscles involved in coarse movement have the largest number, as opposed to muscles which perform delicate movements.

The area of the nerve which lies closest to the muscle cell is known as the synapse. The motor end plate is a specialized area in the muscle cell which fronts the nerve. The synapse and the end plate are separated by a gap (approximately 20 nm) called the synaptic or junctional cleft. The gap is filled with extracellular fluid. As the action potential arrives at the synapse, it causes the neurotransmitter acetylcholine to be released. Acetylcholine then travels across the synaptic cleft and binds selectively to the acetylcholine receptor at the post-synaptic motor end plate of the muscle, causing an action potential to travel through it.

The amount of acetylcholine released following a nerve action potential is far in excess of what is needed to reach the threshold at the end plate. The acetylcholine receptor acts like a switch where it remains closed until the acetylcholine binds to it. It then opens to allow current to pass through it. It closes immediately when the acetylcholine detaches from the receptor.

Acetylcholinesterase

Acetylcholine molecules that do not react with a receptor or are released from the binding site are destroyed almost immediately by the enzyme acetylcholinesterase in the junctional cleft. There are two distinct types of cholinesterase, namely acetyl-cholinesterase (AChE) and butyrylcholinesterase (BChE). They are closely related in molecular structure but differ in their distribution, substrate specificity and functions. AChE is mainly membrane bound, relatively specific for acetylcholine and responsible for its rapid hydrolysis at cholinergic synapses. BChE or pseudo-cholinesterase is relatively non-selective and is found in many tissues. It hydrolyses butyrylcholine more rapidly than acetylcholine as well as other esters such as procaine and succinylcholine.

Conditions That Manifest as Acute Neuromuscular Weakness

There are many conditions that can cause acute muscular weakness or paralysis. In primary neuromuscular disorders, the pathology can occur centrally in the brain down to the peripheral nerves including the neuromuscular junctions, and the muscle itself. Many systemic disorders, as shown in Table 17.1, can also bring about muscle weakness.

Drugs Used to Treat Primary Neuromuscular Diseases

Drugs used to treat primary neuromuscular diseases work at different sites, depending on the pathology. In myasthenia gravis there is a transmission failure in the neuromuscular junction. This is due to an autoimmune response that destroys the nicotinic acetylcholine receptors. Drugs that can enhance cholinergic transmission, such as anticholinesterases, can improve neuromuscular function in this condition. However if the disease is too severe, the number of receptors remaining may become too few to produce an adequate end-plate potential, thus rendering anticholinesterase treatment ineffective. Other treatment modalities for myasthenia gravis involve the use of corticosteroids, intravenous immunoglobulins and plasmapheresis, with the aim of suppressing the amount of circulating immuno-globulins which attack the ACh receptors.

Drugs That Enhance Cholinergic Transmission

Cholinergic transmission can be enhanced either by inhibiting cholinesterase or by increasing acetylcholine release. The peripherally acting anticholinesterase drugs

Table 17.1 Conditions causing muscle weakness or paralysis

Life-threatening central causes of weakness:
Ischemic stroke
Intracerebral hemorrhage
Subarachnoid hemorrhage
Brainstem Stroke
Spinal cord inflammation or compression
Peripheral nerve disease:
Guillain-Barre syndrome (GBS)
Tick paralysis – Dermacentor ticks
Neuromuscular Junction Disease:
Myasthenia gravis
Botulism
Muscle disease:
Alcoholic myopathy
Dermatomyositis & polymyositis
Metabolic Disorders:
Hypoglycemia
Electrolyte abnormalities – potassium, calcium, magnesium and phosphate abnormalities
Endocrine abnormalities:
Thyrotoxic periodic paralysis
Adrenal insufficiency
Hypothyroidism
Infection:
Sepsis
Occult infection – especially in the elderly
Poisoning:
Organophosphate and carbamate poisoning
Carbon monoxide poisoning
Other neurologic causes:
Multiple sclerosis
Hemiplegic migraine
Post-ictal (Todd's) paralysis
Rheumatologic disease:
Systemic lupus erythematosus (SLE)
Rheumatoid arthritis (RA)
Polymyalgia rheumatic
Other medical causes:
Acute coronary syndrome
Anemia
Presyncope
Dehydration or hypovolemia
Medications:
Adverse effects – Beta blockers, diuretics, laxatives, chemotherapeutic agents, isoniazid, opioids and alcohol
Myotoxic effect – glucocorticoid, statins, antimalarial drugs, antipsychotic drugs, colchicines, antiretroviral and cocaine

are mainly used for neuromuscular disorders, whereas the centrally acting anticholinesterase agents have been found to be effective in the treatment of neuropsychiatric disorders such as dementia. There are no clinically useful drugs which increase the release of ACh.

Short-acting Anticholinesterase

These agents are used to diagnose myasthenia gravis (Tensilon test) or to determine the effectiveness of treatment with anticholinesterase therapy

Edrophonium

This is the only short-acting agent available. It is a synthetic quaternary ammonium compound. It binds by a non-covalent bond to AChE at the anionic site, thus competitively preventing acetylcholine binding. It is used mainly for diagnostic purposes as improvement of muscle strength by an anticholinesterase is characteristic of myasthenia gravis. The dose used for Tensilon test to diagnose myasthenia gravis is 2 mg intravenously. Peak effect is attained within 0.8–2 min and the duration of action is only 10 min. The short duration of action is due to rapid reversibility of its binding to AChE and rapid renal elimination.

Medium-duration Anticholinesterases

These are used to reverse the effects of non-depolarizing muscle relaxants following their administration during anesthesia and for the treatment of myasthenia gravis. Drugs in this group include neostigmine and pyridostigmine, which are quaternary ammonium compounds, as well as physostigmine, which is a centrally acting tertiary amine, occurring naturally in the Calabar bean. These drugs are carbamyl esters which bind to the esteratic site of acetylcholinesterase, as does acetylcholine, but the carbamylated enzyme is very much slower to hydrolyze, taking minutes rather than microseconds, leaving more ACh to compete for the post-junctional ACh receptors. Neostigmine is used in the operation theatre to reverse the effects of the non-depolarizing muscle relaxants used to paralyze patients, by increasing the amount of ACh to compete with and replace the muscle relaxants at the receptor sites. Pyridostigmine is used in the treatment of myasthenia gravis where the buildup of ACh is able to sustain the transmission of the nerve impulse across the neuromuscular junction, improving muscle strength.

Neostigmine

This drug is used for the reversal of non-depolarizing neuromuscular blockade as well as in the treatment of myasthenia gravis, paralytic ileus and urinary retention. Neostigmine presents as a 15 mg tablet of neostigmine bromide or as a clear, colorless solution for injection containing 2.5 mg/ml neostigmine methylsulfate. The oral dose for adult is 15–50 mg 2–4 hourly. The intravenous dose for the reversal of non-depolarizing neuromuscular blockade is 0.05–0.07 mg/kg. It should be given with an appropriate anticholinergic agent to minimize cholinergic (muscarinic) effects in the rest of the body. The peak effect when administered intravenously occurs at 7–11 min and a single dose has a duration of 40–60 min.

Pyridostigmine

Pyridostigmine is not used for antagonism of neuromuscular block owing to its slow onset time (more than 16 min). It has a longer duration of action (6 h). Its elimination half-life is 113 min, which makes pyridostigmine the anticholinesterase of choice for myasthenia gravis. The oral dosage is highly individualized, ranging from 60 to 1,500 mg per day. The usual dose is 600 mg per day divided into 5–6 doses. Sustained release formulation is also available with doses ranging from 180 to 540 mg once to twice daily.

Irreversible Anticholinesterases

These are pentavalent phosphorus compounds containing a labile group such as fluoride or an organic group (e.g., parathion and ecothiopate). The serine hydroxyl group of acetylcholinesterase is phosphorylated when the labile group is released. Most of these organophosphate compounds were developed as war gases and pesticides. The inactive phosphorylated enzyme is usually very stable and can cause life-threatening prolonged paralysis. However they are useful agents in the treatment of chronic glaucoma. An example of an irreversible anticholinesterase in clinical use is ecothiopate eye drops in the treatment of chronic glaucoma.

Ecothiopate

This long-acting acetylcholinesterase inhibitor, applied topically, enhances the activity of endogenous acetylcholine, leading to continuous stimulation of the ciliary muscle producing miosis (pupillary constriction). Other effects include potentiation of accommodation and facilitation of aqueous humor outflow, with attendant reduction in intraocular pressure. The onset of action of miosis is 10–30 min. The decrease in intraocular pressure will take 4–8 h to occur.

Malathion, Parathion

These are organophosphate compounds used as insecticides. They are mentioned to aid in the understanding of their effects and the management after accidental ingestion. They are irreversible anticholinesterases which bind to the esteratic site of the AChE enzyme, inactivating it. Upon ingestion, organophosphates increase the ACh concentration in the body to toxic levels causing the patient to suffer the consequences of both muscarinic and nicotinic effects of ACh. These include increased secretions, bradycardia and also muscle weakness.

Clinical Correlation – Organophosphate Poisoning

One of the acute poisoning scenarios frequently encountered in the emergency department is deliberate ingestion of organophosphate compounds. Organophosphate compounds as mentioned earlier are commonly found in pesticides. The clinical presentation exhibits the muscarinic and nicotinic effects of excessive ACh.

Muscarinic signs can be remembered by these mnemonics:

- SLUDGE/BBB – **S**alivation, **L**acrimation, **U**rination, **D**efecation, **G**astric **E**mesis, **B**ronchorrhea, **B**ronchospasm, **B**radycardia

- DUMBELS – **D**efecation, **U**rination, **M**iosis, **B**ronchorrhea/Bronchospasm/ Bradycardia, **E**mesis, **L**acrimation, **S**alivation

Nicotinic effects include fasciculation, tremors, weakness, miosis, and bradycardia. Neurological features include agitation, seizure and coma.

Clinical Correlation – Management of Organophosphate Poisoning

Immediate management includes securing the airway and breathing as well as stabilizing the circulation. Gastric lavage is controversial although some clinicians may perform it in patients who present less than 1 h following ingestion of an organophosphorus agent.

Activated charcoal may be given to patients presenting within 1 h of ingestion of an organophosphorus agent to reduce the absorption. Atropine competes with ACh at muscarinic receptors, preventing cholinergic activation and this may be titrated to the clinical end-point of clearing of respiratory secretions and the cessation of bronchoconstriction.

Since atropine does not bind to nicotinic receptors, it is ineffective in treating muscle weakness. Pralidoxime (2-PAM) and other oximes are cholinesterase reactivating agents that are able to bind to the organophosphate compounds attached to AChE and release it from the enzyme. Pralidoxime should be administered intravenously slowly over 15–30 min as rapid infusion can lead to cardiac or respiratory arrest. The efficacy of pralidoxime is however not proven.

Key Concepts

- Muscle weakness in a patient can arise from a primary neuromuscular disorder or be a manifestation of a systemic illness.
- Most agents used in the management of neuromuscular weakness work at the neuromuscular junction.
- The level of acetylcholine at the neuromuscular junction can be increased by inhibiting its breakdown through the use of acetylcholinesterase.
- Anticholinesterases are used to reverse muscle weakness in myasthenia gravis, in creating miosis, thus reducing intra-ocular pressure in glaucoma management and for reversing muscle relaxants.

Summary

Neuromuscular disorders usually manifest as muscle weakness. The weakness can arise from primary neuromuscular system or be a manifestation of a systemic illness. Few agents are available that act directly on the neuromuscular system to treat muscular weakness. Most of the agents currently available work at the neuromuscular junction to increase acetylcholine levels by inhibiting acetylcholinesterase. Cholinesterase inhibitors are used to treat a few disease conditions such as myasthenia gravis and glaucoma, as well as to reverse the effects of muscle relaxants administered in the operation theatre. Treatment of organophosphate (irreversible anticholinesterase) poisoning is based on the reactivation of the enzyme although the effectiveness is undetermined.

Further Reading

1. Asimos AW. Evaluation of the adult with acute weakness in the emergency department. (Hockberger RS, editor). 2013, July 8. Retrieved from UpToDate: http://www.uptodate.com/contents/evaluation-of-the-adult-with-acute-weakness-in-the-emergency-department. Last accessed 13 June 2014.
2. Eddleston M, Roberts D, Buckley N. Management of severe organophosphorus pesticide poisoning. Crit Care. 2002;6:259.
3. King JM, Hunter JM. Physiology of the neuromuscular junction. Br J Anaesth CEPD Rev. 2002;2(5):129–33.
4. Rang PH, Dale MM, Ritter JM, Moore PK. Cholinergic transmission. In: Pharmacology. 5th ed. Edinburgh/New York: Churchill Livingstone; 2003. p. 136–60.

Chapter 18
Drugs for the Management of Pain

Ramani Vijayan

Abstract Pharmacotherapy is the mainstay in the management of pain, particularly in patients with acute pain. Assessment of the severity and quality of pain is the first step to decide on the most appropriate drug to choose. Acute pain, whether it is somatic or visceral, is mainly nociceptive in character and can be relieved by acetaminophen, non-steroidal anti-inflammatory drugs (NSAIDs), opioids and local anesthetic blocks. The choice of drugs to manage pain depends on the severity of pain. Mild to moderate pain can be relieved with acetaminophen and NSAIDs and weak opioids can be added if pain relief is not adequate. Strong opioids should be used when a patient has severe pain. The intravenous route is used to rapidly relieve severe uncontrolled pain. Acetaminophen and NSAIDs can be added to opioids to enhance pain relief as well as reduce the adverse effects of opioids, which are nausea and vomiting, sedation, pruritus and respiratory depression. Caution must be exercised when using NSAIDs in the elderly, in patients with renal impairment, hypovolemia and bleeding. Analgesic drugs should be given on a regular schedule and patients monitored for efficacy and adverse effects.

Keywords Analgesic drugs • Acetaminophen • NSAIDs • Opioids • Adverse effects of drugs • Mechanism of action of analgesic drugs

Introduction

The mainstay of pain management remains the use of pharmacological agents, particularly for acute pain. Initial assessment of the severity and quality of pain gives us a general idea whether it is nociceptive pain and if it is somatic or visceral in origin.

A large number of medications are available to control pain, but the needs of the patient can only be well served if the provider understands not only how and where the drug works but how to move on to use more appropriate drugs if the initial drugs

R. Vijayan, M.B.B.S., FFARCS, FRCA, FANZCA (✉)
Department of Anesthesiology, Faculty of Medicine, University of Malaya,
50603 Kuala Lumpur, Malaysia
e-mail: ramani@ummc.edu.my

© Springer International Publishing Switzerland 2015
Y.K. Chan et al. (eds.), *Pharmacological Basis of Acute Care*,
DOI 10.1007/978-3-319-10386-0_18

fail. Selection of drugs must also take into consideration the adverse effects that can arise from their use such as nausea and vomiting, gastritis, sedation and respiratory depression.

Pain Pathway and Sites of Drug Action

An understanding of the physiology of pain and the pain pathway is useful for understanding how and where analgesic drugs are effective. Figure 18.1 shows the pain pathway and the sites of action of the various classes of medications that are used to alleviate pain.

- Non-steroidal anti-inflammatory drugs (NSAIDS) and local anesthetics (LA) are effective at the periphery where transduction of pain signals arise
- Local anesthetics are effective in controlling pain all along the afferent axons conducting pain signals until it enters the spinal cord at the dorsal horn. In addition ketamine and α-2 agonists act on transmission at the spinal cord level.
- Tramadol and tricyclic antidepressants (TCA) act on modulating mechanisms.
- Opioids act on the opioid receptors in the brain altering the perception of pain as well on the opioid receptors in the dorsal horn of the spinal cord by inhibiting or reducing the onward transmission of pain signals.

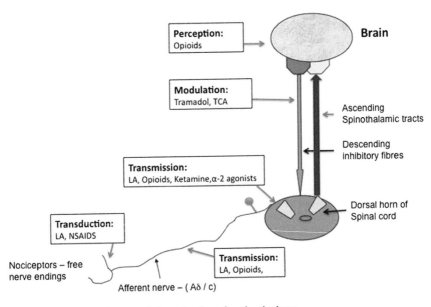

Fig. 18.1 The pain pathway and sites of action of analgesic drugs

Pharmacological Management

First Line of Management

The use of analgesic medications is the first line of treatment. The drugs and choice of route that one uses depends on the severity of pain. Use the Analgesic ladder to choose appropriate drugs (Fig. 18.2).

For example, when a patient is in severe pain (pain score >7/10) the choice of analgesic is an opioid, which should be administered via the intravenous route. The choice of opioid is either morphine or fentanyl.

Intravenous Opioid Administration

Morphine is available in an ampoule in a concentration of 10 mg/ml. It should be diluted in a 10 ml syringe with water/normal saline to a concentration of 1 mg/ml. Rapid relief can be achieved by providing small bolus doses of morphine (0.5–1.0 mg) until the patient is comfortable, i.e. the opioid is titrated until pain relief is achieved. There is a wide range in individual analgesic requirements hence this

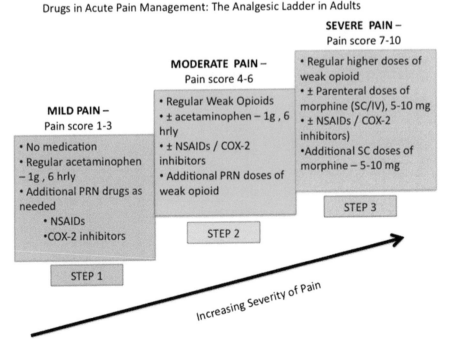

Fig. 18.2 The analgesic ladder

technique provides pain relief adequate to the needs of a patient. Direct intravenous administration bypasses the need for the drug to be absorbed into the systemic circulation either from the subcutaneous tissues (if administered by subcutaneous route) or from the gut (if given orally). It is also immediately available to cross the blood-brain barrier to act on the opioid receptors in the brain.

By giving small bolus doses and waiting for 3–4 min between doses, you can rapidly control severe pain while not causing sedation or respiratory depression. The morphine pain protocol is a useful guide for a beginner (Fig. 18.3).

MORPHINE PAIN PROTOCOL

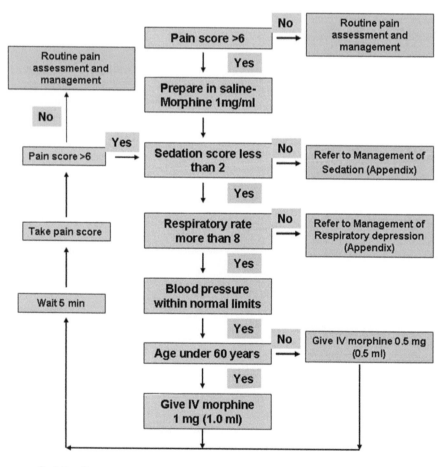

Sedation Score:

0 = Alert, 1 = Mildly drowsy, 2 = Very drowsy but arousable; 3 = Not arousable

Fig. 18.3 Morphine pain protocol

Mechanism of Action of Opioids

Opioids belong to a family of drugs that act on the opioid receptors in the brain and spinal cord to provide analgesia. Opioid receptors were discovered in the 1970s and we currently know that there are three main types of opioid receptors (μ, κ, δ). In the brain they are found in the regions of the sensory cortex, anterior cingular cortex (ACC), insula, periaqueductal gray (PAG) in the midbrain and rostral ventromedial medulla (RVM). In the spinal cord they are found at both the pre-synaptic and post-synaptic regions of dorsal horn neurons.

Opioids are agonists and act via the inhibitory G-proteins of the opioid receptors. Opioids modulate and reduce incoming ascending pain signals traveling to the brain and also activate descending inhibitory pathways, which travel from the midbrain to the spinal cord.

The μ receptor is the main receptor that all the potent opioids act on. Opioids are classified as weak and strong opioids depending on their binding affinity to the μ receptor.

Weak opioids are tramadol and codeine. Tramadol is also known as an atypical opioid as it has a dual mechanism of action. Codeine is a pro-drug, which has to be metabolized in the liver to morphine for its analgesic action. Hence codeine has no analgesic action in patients who cannot metabolize the drug.

Strong or potent opioids used commonly for the relief of severe pain are:

Morphine, pethidine, fentanyl, oxycodone – available for parenteral use (IV/SC)

Morphine and oxycodone are also available in the oral form; both as immediate-release formulations, i.e. aqueous morphine and immediate-release oxycodone capsules (Oxynorm®), and slow/sustained-release tablets, i.e. morphine sulfate (MS Contin®) and oxycodone (Oxycontin®).

Morphine is considered as the gold standard and all other opioids are compared to it, both in terms of potency and lipophilicity.

Pharmacokinetics of Opioids

Table 18.1 shows the pharmacokinetic data of various opioids, which helps us to understand the speed and duration of action, elimination characteristics and the rationale for their use in different situations.

Alfentanil and remifentanil are short-acting opioids that are mainly used by anesthesiologists during surgery. The rapidity of onset of analgesia depends on the percentage of unionized drug in the plasma on intravenous administration. Both drugs have a pKa below the pH of plasma – 7.4, hence highly unionized (90 % and 68 %) and can cross the blood-brain barrier rapidly to produce its analgesic effect. The volume of distribution is also small compared to the other three opioids so they are rapidly eliminated from the plasma accounting for their short duration of action. They are not suitable opioids for use in the wards or in the emergency & accident unit.

Table 18.1 Pharmacokinetics of commonly used opioids

	Morphine	Meperidine	Fentanyl	Alfentanil	Remifentanil
pKa	8.0	8.5	8.4	6.5	7.1
Unionized at pH of 7.4 (%)	23	5	9	90	68
Plasma protein bound (%)	30	40	84	90	70
Terminal half life (hrs)	3	4	3.5	1.6	0.06
Clearance (ml/kg/min)	15-30	8-18	0.8-1.0	4-9	30-40
Volume of distribution (L/kg)	3-5	3-5	3-5	0.4-1.0	0.2-0.3
Relative lipid solubility	1	28	580	90	50

Fentanyl is more rapid in onset compared to morphine; although only 9 % of the drug is unionized in the plasma it penetrates the blood brain barrier more rapidly because of high lipid solubility (see Table 18.1). Fentanyl is available in a 2 ml glass ampoule at a concentration of 50 μg/ml. It should be diluted with water or normal saline in a 10 ml syringe, and boluses of 20 μg of fentanyl should be given intravenously until the patient is more comfortable and the pain score is less than 4/10. There should be at least 3 min between doses.

However, the duration of action of morphine is longer than fentanyl, although the volume of distribution is the same as fentanyl. This is because morphine has the least lipid solubility of all the commonly used opioids. It is cleared less rapidly from the CNS.

Continuation of Analgesia

When severe acute pain has been controlled following an injury, a patient will need ongoing analgesia until a definitive diagnosis is made and appropriate management is undertaken. This plan of action depends on the cause of the onset of acute pain, which can be trauma, acute abdomen or any other painful medical condition. The analgesic regime can include either starting the patient on intravenous patient controlled analgesia (PCA) with morphine or regular doses of subcutaneous (SC) morphine. This can range from 5 to 10 mg, 4–6 hourly.

The dose of morphine that is needed depends on the response of the patient as the analgesic requirement varies individually. The initial dose that one starts with depends on the age of the patient rather than the weight in adult patients. You should start with small doses (5 mg) in older patients.

All opioids are associated with side effects. Around 30 % of patients have nausea/vomiting with opioids as the chemoreceptor trigger zone in the midbrain is close to opioid receptors. Pain score and sedation should be monitored regularly and the subsequent dose of morphine should be omitted if the patient becomes drowsy and difficult to arouse. Sedation is an early indicator of impending respiratory depression. (See Appendix for management of nausea/vomiting, sedation and respiratory depression)

The addition of NSAIDs (traditional or COX-2 specific inhibitors) can reduce opioid requirements (opioid sparing) and can reduce the incidence of nausea/vomiting. In addition because it has a different mechanism of action (see below) it enhances the quality of pain relief. This is often referred to as multi-modal analgesia. Parenteral opioids can be changed to oral opioids, when a patient is able to take oral medications and/or the patient is to be discharged. Oral tramadol is the drug of choice. Tramadol is known as an atypical opioid with a dual mechanism of action. It has weak opioid properties at the μ opioid receptor and inhibits the re-uptake of noradrenaline and serotonin at the descending inhibitory tracts. It is thus an effective analgesic drug with less sedation and respiratory depression when compared to strong opioids.

Mechanism of Action of NSAIDS

Tissue injury results in an outpouring of inflammatory mediators, with prostaglandins being one of the important mediators. These mediators activate nociceptors that are on the free nerve endings of 'Aδ' and 'c' afferent fibres and set up action potentials, which are conducted via these afferent fibres to the dorsal horn of the spinal cord where they form synaptic connections with second order neurons.

Injury to cells releases membrane phospholipids that are hydrolyzed to arachidonic acid by phospholipase A_2. The subsequent metabolism of arachidonic acid in the tissues is responsible for the production of some of the inflammatory mediators (Fig. 18.4). Arachidonic acid is metabolized by two enzymes, which are induced in the presence of tissue injury, cyclooxygenase-2 (COX-2) and lipooxygenase. The action of COX-2 results in the production of prostaglandins (PGE3 and PGF2), which are responsible for pain and inflammation and thromboxane that is important for platelet aggregation. The action of lipooxygenase on arachidonic acid results in leucotrienes, which are also important mediators of pain and inflammation.

NSAIDs are a mixed group of compounds that reduce pain and swelling by inhibiting COX-2 enzyme thereby reducing prostaglandins, which are powerful mediators in the pain signaling pathway. They have no action on other inflammatory mediators such as bradykinin, serotonin or leucotrienes. Hence they are useful for moderate pain when there is an inflammatory component. They can be classified into the traditional NSAIDs (tNSAIDs) and the selective COX-2 inhibitors also known as coxibs.

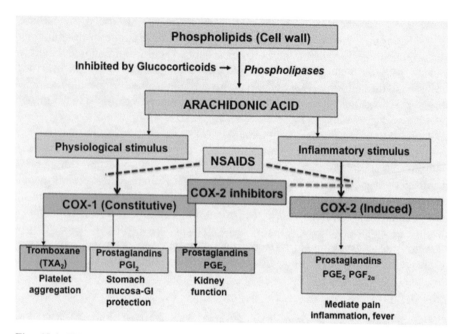

Fig. 18.4 This shows the two isoforms of the cyclooxygenase enzymes involved in the metabolism of arachidonic acid (With permission from Springer; Inflamm Res 1995;44(1):1–10)

The COX enzyme exists in two isoforms (Fig. 18.4). COX-1, which is the constitutive enzyme found in the gastric mucosa and platelets, is important for gastric mucosal protection and platelet aggregation. The COX-2 enzyme is induced in the presence of tissue injury but is also an important constitutive enzyme in the kidney, being important for the maintenance of renal blood flow and glomerular filtration.

Traditional NSAIDs such as diclofenac, mefenamic acid, naproxen and ibuprofen are non-selective and inhibit both COX-1 and COX-2 enzymes. They therefore have the potential to reduce the protective action of COX-1 on the gastric mucosa and reduce platelet aggregation resulting in gastritis and GI bleeding. The selective COX-2 inhibitors such as celecoxib and etoricoxib have a reduced incidence of adverse GI side effects. However, both types of NSAIDs can affect renal blood flow and should be avoided in the presence of hypovolemia and renal impairment.

Major Side Effects of NSAIDS

(a) Gastric irritation / ulceration (Less with COX-2 inhibitors)
(b) Anti-platelet effect leading to bleeding (Not with COX-2 inhibitors)
(c) Reduced renal blood flow with both NSAIDs and COX-2 inhibitors

1. Avoid in patients with hypovolemia and hypotension
2. Long-term use may result in renal impairment

(d) Cardiovascular risk – increased risk of stroke and myocardial infarction with long-term use (both drugs)

(e) Allergic reaction in patients who are sensitive (cross-sensitivity between different NSAIDs is possible)

Clinical Correlation – Multi-modal Analgesia

The use of different classes of analgesic drugs with different mechanisms of action to control pain is acceptable and is known as multi-modal analgesia. This is an acceptable practice when analgesic drugs used act on different sites of the pain pathway. When NSAIDs or COX-2 specific inhibitors are added to opioids in a patient with moderately severe pain, the quality of analgesia is enhanced. There is also opioid sparing, i.e., the patient needs less opioids for equivalent analgesia and thus the adverse effects of opioids can be reduced. However, one should not use two drugs with the same mechanism of action, as that can lead to toxicity.

Other Considerations in Managing the Patient in Pain

Pain is both a sensory as well as an emotional experience, hence it is important to assess the patient's level of anxiety with regards to pain and level of disability and a patient in pain should be attended to promptly and reassured that acute pain will be controlled

Other Analgesic Drugs

Parecoxib

This is an injectable COX-2 specific inhibitor. It is effective in controlling moderate pain and useful when a patient's oral intake is restricted or the patient is vomiting. It has an opioid sparing effect. It is available in a vial as a powder and should be reconstituted with 2 ml of normal saline before administration. The dose for adults is 40 mg twice a day. In Malaysia it is licensed only for 2 days and this should not be exceeded. Contraindications to its use are similar to other NSAIDs and coxibs as well as a history of allergy to sulfa drugs.

Acetaminophen

Acetaminophen can be used along with NSAIDs and opioids to enhance analgesia. The mechanism of its analgesic action is unclear and is believed to be at the level of the brain and spinal cord. The maximum dose of acetaminophen is 1 g every 6 h and 4 g/day should not be exceeded in adult patients. Reduce the dose to 500 mg every 6 h in patients who are less than 50 kg in weight and/or elderly frail patients. It should be avoided in patients with liver impairment.

Ketamine

This drug in small doses can provide analgesia and is a useful drug to relieve acute pain in the presence of hypotension and hypovolaemia. It should be given in 10 mg boluses to a maximum of 30 mg. Ketamine is an NMDA antagonist and exerts its action centrally. It can increase cerebral blood flow and is contraindicated in patients with head injury.

Oxycodone

Oxycodone is a synthetic opioid that has been available in Malaysia for the last 2 years. (The injectable immediate release form is known as Oxynorm). For the relief of severe pain it can be treated as similar to morphine. Oxycodone is also available in the oral form and its bioavailability is 80 % and patients can be given PR (prolonged release) oxycodone or IR (immediate release) oxycodone for out-patient use for a few days. It is metabolized in the liver to inactive metabolites. In contrast, oral morphine has a bioavailability of 30 %, is metabolized in the liver and has an active metabolite. Both oral morphine and oxycodone are used commonly for the relief of acute cancer pain.

Meperidine

It is no longer recommended for postoperative pain relief and for chronic or recurrent pain conditions because of its active metabolite, norpethidine. This can accumulate and cause convulsions with prolonged use and in patients with renal impairment. Meperidine is also thought to have more potential for the development of addiction when compared to other opioids, when administered for recurrent acute on chronic pain.

Key Concepts

- When a patient is in pain, the patient should be initially assessed to determine its severity and type, so as to be able to provide the appropriate medications for pain relief.
- Opioids are the mainstay for the management of acute pain and it should be administered via the intravenous route if pain is very severe.
- When pain is under control, analgesic drugs should be given regularly for a few days depending on the severity and the route of administration depends on the situation; if the patient is to be admitted or managed as an outpatient.
- Non-opioid analgesic drugs such as NSAIDs/COX-2 specific inhibitors/ acet-aminophen can be combined with opioids to reduce the requirement of opioids for optimum pain relief, so that the side effects of opioids can be minimized.
- Pain is both a sensory as well as an emotional experience, hence it is important to assess the patient's level of anxiety with regards to pain and level of disability and a patient in pain should be attended to promptly and reassured that acute pain will be controlled.

Summary

Assessment of pain is an important first step in the management of pain. The use of medication is the first line of treatment. An understanding of the pharmacology of individual drugs is important so that we can select the one that best meets the patient's needs. Multi-modal analgesia where drugs with different mechanisms of action are combined, results in better quality analgesia. Non-pharmacological considerations are also important in pain management.

Appendix

Management of Adverse Side Effects

1. *Nausea and Vomiting*
   - Nausea and vomiting occur in about one third of patients given opioids, both potent and weak opioids
   - It is necessary to treat this side effect with anti-emetics while continuing to prescribe opioids for pain relief. Do not stop the opioid due to nausea and vomiting. (It is a side effect and not an allergic reaction)
   - First line anti-emetic is:

1. Metoclopramide (10–20 mg; oral/IV/SC) give one stat dose and repeat as necessary 6–8 hourly
2. If not effective and patient continues to have nausea/vomiting

    (a) Ondansetron 4 mg IV – give one dose and repeat if necessary 8 hourly OR
    (b) Granisetron 1 mg IV – give one dose and repeat if necessary 8 hourly OR
    (c) If the above are not available, dexamethasone 4 mg IV / haloperidol 1.5 mg (oral) can be used.

- If a patient has persistent vomiting despite adequate anti-emetic therapy, change to another opioid or look for other causes of vomiting (surgical causes).

2. *Sedation*

- Sedation can occur, particularly with the first dose of opioid.
- A sedation score of 2 is an early sign of respiratory depression and should be taken seriously. Regular pain and sedation scoring is important.
- Oxygen should be administered by facemask and the patient monitored while the next dose of opioid should be withheld.
- When the patient is alert, opioids can be resumed for pain relief, at a lower dose and at longer intervals. (Look for hepatic and renal impairment)

3. *Respiratory Depression*

- When the sedation score is 3 (patient not arousable) OR the sedation score is 2 (difficult to arouse) and the respiratory rate is <8 breaths/min.
- It can occur with overdose of opioids
- However, it is uncommon and the risk of respiratory depression is minimal if strong opioids are titrated to effect and only used for the relief of pain.

Further Reading

1. Graham GG, Scott KT. Mechanism of action of paracetamol. Am J Ther. 2005;12(1):46–5.
2. Inturrisi CE. Clinical pharmacology of opioids for pain. Clin J Pain. 2002;18(4 Suppl):S3–13.
3. Kohr R, Dureux M. Ketamine: teaching an old drug new tricks. Anesth Analg. 1998;87 (5):1186–93.
4. Pathan H, Williams J. Basic opioid pharmacology-an update. Br J Pain. 2012;6(1):11–6. Available from: http://bjp.sagepub.com/content/6/1/11.full.pdf. Last accessed 23rd June 2014.
5. Raffa RB, Friderichs E, Reimann W, Shank RP, Codd EE, Vaught JL. Opioid and non-opioid components independently contribute to the mechanism of action of tramadol, an atypical opioid analgesic. J Pharmacol Exp Ther. 1992;260(1):275–85.
6. Vane JR, Bottling RM. New insights into the mode of action of anti-inflammatory drugs. Inflamm Res. 1995;44(1):1–10.

Part III
Pharmacology in Special Circumstances

Chapter 19
The Patient in Pain

Ramani Vijayan

Abstract Pain has been defined as an unpleasant sensory and emotional experience associated with actual or potential tissue damage. It is one of the common reasons for anyone to seek medical help and occurs in a wide variety of conditions. It can be either acute or chronic. Acute pain is awareness of noxious signals from recently damaged tissues (somatic or visceral) and is regarded as a symptom or warning signal of some underlying pathology. Chronic pain is defined as pain which persists for longer than 3 months. It can occur in chronic inflammatory conditions, with nerve injury or even when there is no obvious pathology and is sometimes considered as a disease of the nervous system. Unrelieved pain leads to unnecessary suffering and may have deleterious effects in vulnerable populations, while negatively impacting on the quality of life. To adequately manage pain, it is important to assess the patient using a detailed pain history to determine if it is acute or chronic, the underlying pathophysiology, factors that aggravate/relieve pain and the impact of pain on the patient's life. This will help the practitioner formulate a plan to treat the patient holistically using appropriate drugs and take appropriate non-pharmacological measures to alleviate pain and suffering.

Keywords Acute pain • Chronic pain • Assessment of pain • Pathophysiology of pain • Suffering • Quality of life

Introduction

Pain is the most common reason for people to seek health care, and as a presenting complaint, it accounts for more than two-thirds of visits to the emergency department. Major categories of acutely painful conditions include trauma to the musculoskeletal system, postoperative pain, renal/abdominal colic, chest pain, headache and pain secondary to upper-respiratory infections.

Acute pain is an awareness of noxious signals from recently damaged tissues and is regarded as a symptom/warning signal of some underlying pathology. It is

R. Vijayan, M.B.B.S., FFARCS, FRCA, FANZCA (✉)
Department of Anesthesiology, Faculty of Medicine, University of Malaya,
50603 Kuala Lumpur, Malaysia
e-mail: ramani@ummc.edu.my

© Springer International Publishing Switzerland 2015
Y.K. Chan et al. (eds.), *Pharmacological Basis of Acute Care*,
DOI 10.1007/978-3-319-10386-0_19

therefore essential for the survival of the species. In ancient times, with no real recourse to any remedies, the presence of pain allowed one to rest the painful part to allow healing.

In modern times, acute pain is still a very useful signal. It allows early referral to health care practitioners. However, once in a controlled medical environment, such as the emergency department, it no longer serves any useful purpose and needs to be controlled as soon as possible. Pain leads to suffering, particularly in vulnerable populations such as the elderly, and those with co-morbidities like ischemic heart disease and chronic respiratory disease, and it can have deleterious effects.

Assessment of Patient in Pain

Management of pain begins with assessment of the patient in pain. This would include a preliminary survey of the patient's general condition. An assessment of a patient's vital signs is essential to ensure cardiovascular and respiratory stability. For example, any instability such as hypotension associated with trauma and blood loss should be dealt with urgently.

A detailed pain history should follow with regards to a patient's main complaint, which would include:

(A) *Onset and duration of pain*

   (a) Was it sudden and associated with any trauma?
   (b) Was the patient unwell when the pain started?
   (c) How long has the patient been having pain?

(B) *Site of pain*

   (a) Is the pain associated with the site of injury?
   (b) Is it abdominal pain?
   (c) Is it chest pain?

(C) Assessment of the *severity of pain*

   (a) Ask the patient to indicate the severity of pain. The 11-point numeric rating scale, where 0 is no pain and 10 is the worst pain imaginable is the pain assessment tool that is commonly used. This will give the attending physician a general idea about the level of pain that the patient is suffering from; whether it is mild, moderate or severe pain and this will help decide the most appropriate analgesic medication.

(D) Assessment of the quality of pain

   (a) Acute pain is usually described as throbbing, stabbing or sharp if it is somatic in origin and dull, aching or constricting if it has a visceral component. Visceral pain is often difficult to localize and can be referred to a somatic area. For example, pain associated with myocardial infarction is referred to the left arm, shoulder or jaw.

(b) If the description of the pain includes words such as intermittent electric shock like or burning, then one should suspect that there is also injury to a nerve.

(E) Factors that aggravate and alleviate pain

(a) Is the pain worse with movement?
(b) Does elevation make it less painful?

Assessment of the severity and quality of pain will provide an indication of whether the pain is nociceptive (or inflammatory) in nature and if there is an element of neuropathic pain. Acute pain is mainly nociceptive pain when the signal transmitting system is normal and this type of pain can be controlled with drugs such as opioids, NSAIDs (traditional NSAIDs or COX-2 selective inhibitors), acetaminophen and nerve blocking techniques with local anesthetic drugs. The control of neuropathic pain includes the use of drugs such as anticonvulsants and antidepressants.

Immediate Non-pharmacological Management

Immediate management should include non-pharmacological measures such as reassurance, immobilization of the site of injury, administration of oxygen in the presence of chest pain and cold compression until pharmacological management is instituted.

Pain as an Emotional Experience

It is also important to remember that pain is both an unpleasant sensory and emotional experience, and there are several factors other than the severity of injury that can enhance a patient's perception of pain. The level of anxiety, the meaning of the injury or disease condition, the level of information provided to the patient, and the absence or presence of caregivers/relatives are some of the factors that can affect a patient's pain. Thus, although drug therapy is important, these other factors should also be taken into account when managing a patient in pain.

Key Concepts

- Management of pain begins with assessment of pain.
- It is important to know that acute pain is usually nociceptive pain, but there may be a neuropathic component.

- Both non-pharmacological and pharmacological methods should be used in the relief of pain.
- The choice of drugs used to control acute pain depends on the severity of pain.

Summary

Pain is a common complaint that leads a patient to seek help. Unrelieved pain results in unnecessary suffering and can have deleterious effects in vulnerable populations. Assessment of pain, which includes a detailed pain history, is the first step towards management. This helps to determine its severity and pathophysiology so that a rational plan of management can be instituted.

Further Reading

1. Breivik H, Borchgrevink PC, Allen SM, Rosseland LA, Romundstad L, Breivik Hals EK, et al. Assessment of pain. Br J Anaesth. 2008;101(1):17–24.
2. Edwards RR, Berde CB. Pain assessment (Chapter 5). In: Benzon HT, Raja SN, Liu SS, Fishman SM, Cohen SP, editors. Essentials of pain medicine and regional anesthesia. 3rd ed. Philadelphia: Saunders, an imprint of Elsevier; 2011.
3. Fink R. Pain assessment: the corner stone to optimal pain management. Proc (Bayl Univ Med Cent). 2000;13(3):236–9.
4. Kishner S. Pain assessment. http://emedicine.medscape.com/article/1948069-overview. Last accessed 13 June 2014.
5. Merskey H, Loeser JD, Dubner R. The paths of pain, editors. Seattle: IASP Press; 2005.
6. Vazirani J, Knott JC. Mandatory pain scoring at triage reduces time to analgesia. Ann Emerg Med. 2012;59(2):134–8.

Chapter 20
The Pediatric Patient

Lucy Chan

Abstract The administration of drugs to the pediatric population should be safe and effective with minimum adverse events. Clinicians who are involved in their management should have "bench" knowledge of the pharmacokinetics and pharmacodynamics of drugs before application at the bedside. Unfortunately, the availability of evidence-based pharmaco-therapeutics in children lags behind the vast volume of clinical information that is published for drugs administered to adults. This lack of scientific basis for many drugs is particularly noted in the young infants, neonates and critically ill children. Doses for administration are derived from weight-based calculations from studies carried out in adults. This may not be suitable for all children as they belong to a heterogeneous group. The changes in the newborn as they grow into adults are seen as a dynamic spectrum in psychological, physiological and anatomical maturation. Understanding the physiological aspect in every child is especially important due to age-related differences in body components. This natural process of growth and development underlies the challenges in clinical pharmacology in children, requiring consideration of dose adjustment when drugs are given. Assessment of the child who is acutely ill will lead to diagnosis and decision-making for drug selection to manage pathophysiological states such as shock, hypoxia and poor perfusion. The further evaluation of the level of body organ function will guide clinicians to modify doses or choice of drugs. In acute care, many drugs may be given simultaneously and it is evident that interactions and toxicities may arise through inappropriate drug regimens. Therapeutic drug monitoring may be particularly useful in providing additional information for drugs with a low therapeutic index.

Keywords Absorption • Distribution • Metabolism • Elimination • Drug interactions • Pharmacokinetics • Pharmacodynamics

L. Chan, M.B.B.S., FANZCA (✉)
Department of Anesthesiology, Faculty of Medicine, University of Malaya,
50603 Kuala Lumpur, Malaysia
e-mail: lucy@um.edu.my

© Springer International Publishing Switzerland 2015
Y.K. Chan et al. (eds.), *Pharmacological Basis of Acute Care*,
DOI 10.1007/978-3-319-10386-0_20

Introduction

It is not unusual that children in the acute care setting receive many drugs that require meticulous attention to dose calculations and monitoring of desired and undesired effects. Registered drugs for children are few and about 70 % of pre-scribed drugs in children are not evidence-based. As the majority of clinical trials are performed in adults, recommendations for pediatric dosing represent adjustment of the adult dosage. Under such circumstances, many drugs may be considered off-label and unlicensed for children.

A good understanding of the principles of pharmacokinetics, pharmacodynamics and drug interactions as they apply to the pediatric population will allow care providers to make the correct adjustment in the administration of drugs to meet the needs of their patients.

Pharmacokinetics

For children, important age-related differences alter the four physiological pro-cesses of absorption, distribution, metabolism and excretion which all affect the determination of the correct dose to be administered to an individual child. Extrap-olating dose calculations from studies done in adults to the child exposes the child to potentially toxic blood or tissue drug levels (Fig. 20.1). Most care providers may be unaware of this issue, but it remains a huge concern for dose determination in the pediatric population.

Absorption

Oral medication for a child who requires acute care (e.g., in shock) is uncommon because there is decreased absorption and delayed drug response from reduced perfusion to the gastrointestinal tract. Nevertheless, it is an easy and effective route for many drugs. Enzymes present in the liver and gastrointestinal tract can metab-olize drugs before they reach the systemic circulation. This is known as the "first-pass effect". When a drug has a significant first-pass effect, a smaller quantity (less than 100 %) of the orally administered dose reaches the site of action. Thus, the drug is said to have low bioavailability. Bioavailability becomes 100 % for intra-venously administered drugs because the whole dose is given into the systemic circulation. Numerous factors influence absorption via the oral route: gastric emp-tying, gastric acidity, intestinal motility, bacterial colonization and enzymatic function (see Chap. 3). This is especially important in the first 2 years of life.

In neonates, gastric emptying is delayed which is reflected in a longer interval to reach a lower peak plasma concentration. Gastric acidity is important for drug

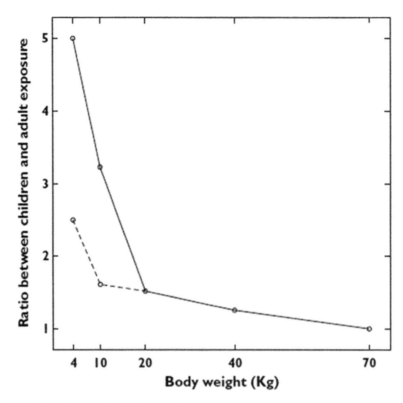

Fig. 20.1 Blood levels of a child exposed to doses of chloramphenicol calculated according to body weight (*solid line*) and blood levels in the child exposed to doses adjusted downwards to avoid the grey baby syndrome (*dashed line*) (Permission from John Wiley and Sons; British Journal of Pharmacology 2010;70(4):597–603)

ionization and absorption. Gastric acidity is decreased in neonates. The more alkaline environment favors the absorption of drugs that are destroyed in acidic media (such as ampicillin) but reduces the uptake of acidic drugs (such as phenobarbitone).

The rectal route for drug administration is undesirable in an acute setting because of the slow and unpredictable absorption of drugs. Despite this, it is an effective method to sedate children with diazepam rectal suppositories. Similarly, intramuscular (IM) injection of medication in children is discouraged. Unpredictable absorption may arise in a critically ill child with hypotension and reduced perfusion to the muscles. Besides, IM injection is painful.

In older children, fentanyl and clonidine transdermal patches offer another route for drug administration. Absorption can be increased in neonates and preterm babies because of the thin skin and better hydration (larger surface area versus body weight). Caution is necessary as there are reports of toxicities following the topical application of medications (e.g., with epinephrine, salicylic acid).

Medications that reach the systemic circulation are, to a certain extent, protein-bound. Infants have decreased protein binding of drugs. Several factors are responsible for this: fetal albumin has decreased binding affinity, reduced amounts of binding proteins (e.g., albumin) are available and there is competition by other agents for the protein binding sites, such as bilirubin or hormones that have crossed the placenta.

Diseases that contribute to hypoalbuminemia, acidosis and hyperbilirubinemia may reduce the protein binding leading to more unbound drug to occupy active sites. Phenytoin is highly protein bound. Clinicians need to identify any altered protein binding so that a measurement of the free or unbound concentration can be determined to better assess therapy.

The majority of medications are administered intravenously to acutely sick children. The pediatric daily maintenance fluid volume averages 1,500 ml/m^2/day. It is necessary to plan an intravenous fluid management strategy, taking into account the rate of infusion and the small volumes required. High concentrations of drugs have adverse effects (e.g., phlebitis) on the smaller veins. Clinicians are encouraged to follow recommendations regarding the maximum concentration and the infusion regime for potent drugs with a short half-life such as norepinephrine. Dilution of drugs should be double-checked by a senior staff before the continuous intravenous infusion of medication is commenced, e.g., mcg/kg/min. Children with cardiac or renal disorders, who are managed on low fluid regime, require attention to fluid balance.

Distribution

Dug distribution following administration into the systemic circulation is affected by physicochemical characteristics of the drugs and the pathophysiological status of the child. In particular, age-related altered body composition (e.g., in preterm neonate, obese child) and the dynamic changes in the critical clinical condition (e.g., shock, tissue perfusion), are special factors that should be considered (see Chap. 4).

When expressed as a percentage of body weight, total body water (TBW) is greater in the newborn and decreases with increasing age (full-term newborn – 75 %, 3 month old – 60 % and adult – 55 %). Extracellular water as a percentage of body weight is 50 % in premature infants, 35 % in infants and 19 % in adults. The higher TBW and extracellular water in infants result in a large volume of distribution (V_d) for water-soluble drugs. V_d is important for calculating the loading dose (first dose), where

$$\text{Loading dose} = V_d \times \text{target plasma concentration}$$

For example, the loading dose of the water-soluble drug gentamicin is larger in infants versus older children. In contrast, neonates require a smaller loading dose of fat soluble drugs (e.g., diazepam).

The differing proportions of fat in the various categories of children need to be taken into consideration. A preterm has 1–2 % fat tissue versus 15 % in a term newborn.

Metabolism

The liver is the major organ for drug metabolism. The process of metabolism inactivates lipid-soluble drugs to water-soluble compounds (polar) for excretion. Metabolized compounds can be an active metabolite (e.g., normeperidine) that contributes to pharmacological effect as well as to toxicities when it accumulates in children with poor renal function.

Metabolism involves Phase 1 and Phase II reactions (see Chap. 5). In Phase 1 there are alterations of molecular structures of drugs (e.g., oxidation, reduction), while Phase 2 involves synthetic processes, such as conjugation (e.g., with glucuronide or sulfate). Both phases are significantly decreased in the newborn. Phase 1 reactions are about 50 % of adult values and this affects the oxidation of some medications (e.g., diazepam, phenytoin). However, by 1 year of age, the oxidative bioactivity is more than double the adult value. The high clearance requires a higher maintenance dose. On this basis, the recommended maintenance dose of phenytoin is 3–4 mg/kg/day during the neonatal period but is 6–8 mg/kg/day throughout early childhood (1–5 years).

Glycine conjugation is required in the metabolism of preservatives such as benzyl alcohol and benzoic acid that are present in many diluents and solutions. In newborns, glycine conjugation is decreased and only matures by about 2 months of age. A potentially fatal syndrome in neonates known as "gasping syndrome" (severe metabolic acidosis, multiple organ system failure) is triggered by excess benzyl alcohol and benzoic acid resulting from this reduced metabolism. It is important that intravenous drugs and diluents are preservative-free for children so that toxicity from these preservatives do not negatively affect the outcome of medical care.

Drug metabolism differs from child to child because each child's metabolic pathway depends on the isoforms and amounts of enzymes expressed. Whilst age-related dose adjustment is relevant, disease states can increase (induce) or decrease (inhibit) liver enzymes. In the acute setting of hypoxia and shock, the child may also have decreased hepatic enzyme activity.

Elimination (Metabolism and Excretion)

Clearance (CL) of a drug describes how efficiently the body removes or eliminates the drug. In pediatric pharmacology, CL is expressed as ml/min/kg or ml/min/m^2. The latter unit is often used as it provides a better measure of clearance in children of different ages and body size. It is necessary to know the clearance of a drug to determine the maintenance dose in multiple dosing:

Maintenance dose rate $=$ CL $\times$ target drug concentration (therapeutic drug concentration).

Both CL and V_d of a drug determine the half-life ($t_{1/2}$) of the drug in the body. The time taken for a drug concentration to decline to half its initial value is known as the half-life of the drug. $T_{1/2}$ may not be a good indicator of elimination but it is useful in the calculation of appropriate dosing interval. The aim of pharmacotherapy is to keep the active drug at therapeutic level constant or at steady-state (see Chap. 6) for the duration of therapy. The time taken for a drug to reach a plateau or steady-state is 4–5 times the $t_{1/2}$ of the drug. Drugs with long $t_{1/2}$ such as 24 h, will take several days to reach steady-state. Phenytoin is an example of such a drug ($t_{1/2} = 22$ h). When the drug concentration is small, phenytoin follows first order (linear) elimination. At this point, the elimination is proportional to the drug concentration. When the concentration reaches the therapeutic level, saturation of metabolizing enzymes has occurred and if a further dose is administered, toxicity may occur. This is referred to as non-linear (zero-order) kinetics where elimination is a rate-constant irrespective of dose. Therapeutic drug monitoring of serum phenytoin is essential to avoid accumulation of drug and toxicity.

Drug excretion relies mainly on the kidneys. Renal function matures with age. In the infants, drug selection or dose rate adjustments are necessary for drugs that are eliminated by the kidneys. In acute care, appropriate individualized drug regimens should consider the possibility of decreased glomerular filtration and altered CL associated with shock or impaired renal function.

Pharmacodynamics

Although a great deal of information is published regarding pharmacokinetic (PK) changes during maturation of the newborn to early childhood and adolescence, there is limited knowledge of pharmacodynamics (PD) in children. Understanding PD enables clinicians to know about pharmacologic effects and toxicities of drugs.

Therapeutic index (TI) is the ratio between the toxic dose and the effective dose, and describes the safety profile of a drug. Drugs with low TI have a narrow margin of safety (e.g., digoxin, theophylline, phenytoin) and require therapeutic monitoring of drug levels. This is due to the small window for therapeutic dosing. In contrast, penicillin has a large TI.

Studies have shown that drug concentrations at a specific time relate to peak clinical effect and such information enables clinicians to use drugs effectively and safely. Details are available in standard pediatric textbooks.

Drug Interactions

The basis of drug interaction is either pharmacokinetic or pharmacodynamic in nature (see Chap. 10). Drug interaction is more likely to have an adverse outcome in children as their compensatory reserves are less. Modification of pharmacologic effects can occur when two incompatible drugs are administered, sometimes with disastrous outcome.

Physicochemical incompatibilities may exist between drugs. For instance, precipitation can occur when calcium is administered intravenously via the same line that has just been used to infuse ceftriaxone. The resulting severe kidney and lung injuries may be fatal in preterm and term neonates.

Drug interactions can occur at the receptor level. For example, an asthmatic child who is administered propranolol (beta-adrenoceptor antagonist) and albuterol (beta-adrenoceptor agonist), will clinically develop bronchoconstriction. The bronchodilatation effect of albuterol is blocked by propranolol.

When several drugs are administered together, the possibility of drug-drug interactions in liver metabolism can lead to unwanted responses. Cytochrome P-450 enzymes play a major role in the metabolism of drugs. Enzyme inhibition by a drug can cause significant increase in plasma levels and toxicity of another drug that is metabolized by the liver. For example, the interaction between erythromycin and theophylline during metabolism can result in theophylline toxicity because erythromycin is an enzyme inhibitor. Similarly, phenytoin is an enzyme inducer and can significantly reduce the efficacy of oral anticoagulants.

Key Concepts

- Pharmacologic characteristics of a drug must be considered for drug selection and dosing guidelines in children under acute care.
- Pharmacokinetic and pharmacodynamic characteristics are influenced by age, body size, organ maturation and acute clinical conditions such as hypoxia, shock, dehydration and asphyxia.
- Potent medications are often given as a continuous infusion and therapeutic drug monitoring is desirable for drugs with narrow therapeutic index.
- Possible drug interactions have to be considered as many drugs are administered to a child in acute care.

Summary

Therapeutic challenges can be encountered in the acute care of children because each drug has its own PK and PD profiles. The pediatric group is a heterogeneous population that is governed by physiologic processes and body systems that are still undergoing development and maturation. Changes in weight, size and body composition occur from birth to childhood and impact strongly on the PK and PD of drugs, ultimately influencing rational drug recommendations in pediatrics. In the acute care setting, multiple drugs are often necessary for management and there is great potential for drug–drug interactions. Extreme caution has to be exercised when managing vulnerable groups (e.g., preterm or neonates) and they may be subjected to more harm than benefit from lack of understanding and unmonitored drug therapy. Acutely sick infants and individuals with organ dysfunction (e.g., asphyxiated child with renal failure) require special attention in pharmacologic interventions.

Further Reading

1. Anderson GD, Lynn AM. Optimizing pediatric dosing: a developmental pharmacologic approach. Pharmacotherapy. 2009;29(6):680–90.
2. Cella M, Knibble C, Danhof M, Pasqua OD. What is the right dose for children? Br J Clin Pharmacol. 2010;70(4):597–603.
3. Downes KJ, Hahn A, Wiles J, Courter JD, Vinks AA. Dose optimization of antibiotics in children: application of pharmacokinetics/pharmacodynamics in paediatrics. Int J Antimicrob Agents. 2014;43(3):223–30.
4. Gorman RL. The march toward rational therapeutics in children. Pediatr Infect Dis J. 2003;22(12):1119–23.
5. Johnson TN, Thomson AH. Design of pharmacokinetic-pharmacodynamics (PK-PD) studies in children: a workshop for health professions involved in pediatric drug research. Paediatr Perinat Drug Ther. 2006;7(1):10–4.
6. Rodriquez W, Selen A, Avant D, Chaurasia C, Crescenzi T, Gieser G, Di Giacinto J, Huang SM, Lee P, Mathis L, Murphy D, Murphy S, Roberts R, Sachs HC, Suarez S, Tandon V, Uppoor RS. Improving pediatric dosing through pediatric initiatives: what we have learned. Pediatrics. 2008;121(3):530–9.

Chapter 21
The Elderly Patient

Pui San Loh

Abstract Many more elderly individuals will present for treatment in acute care now than in the past as the proportion of patients in the geriatric age group increases. Physiological changes in the elderly affect the way their bodies handle and react to various drugs, and this needs to be kept in mind during the management of these patients in acute care. Some pharmacological changes can be predicted while many others remain unknown simply because large randomized controlled trials are difficult to perform in this age group. Many who present in the acute setting may already have medical co-morbidities and be on multiple medications which will increase their risks for drug-drug interactions and associated adverse effects. The pharmacokinetic changes associated with ageing include changes in absorption due to decreased blood flow, altered volume of distribution depending on lipid solubility or water solubility, decreased metabolism and excretion. It is imperative that drug administration in this category of patients is titrated to meet the patients' needs.

Keywords Elderly • Geriatric • Titration of drugs • Decreased metabolism • Multiple medications • Drug interactions

Introduction

Aging is part of a normal physiological process that transforms many systems in our body. The geriatric group can be difficult to classify by age but it is generally accepted to be 65 years and above. Statistics have shown that our patient population is aging fast and growing in number.

This group of patients tend to have multiple pre-existing medical morbidities which naturally put them at risk of requiring acute and long term medical management. Practitioners who provide acute care to the elderly must consider and understand many pharmacological issues peculiar to them. As with the group of

P.S. Loh, M.B.B.S., MMed Anaes, FANZCA (✉)
Department of Anesthesiology, Faculty of Medicine, University of Malaya,
50603 Kuala Lumpur, Malaysia
e-mail: lohps@um.edu.my

© Springer International Publishing Switzerland 2015
Y.K. Chan et al. (eds.), *Pharmacological Basis of Acute Care*,
DOI 10.1007/978-3-319-10386-0_21

patients in the other extreme of age, the paediatric group, the pharmacokinetics and pharmacodynamics of many drugs in the geriatric population have not been studied sufficiently and their dosing requirements or actions remain predictions based on large randomized trials in a much younger adult population. Age related changes in pharmacokinetics, pharmacodynamics and adverse drug reactions including drug-drug interactions are important issues in the care of this category of patients.

Pharmacokinetic Changes in the Elderly

With aging, the body physiology changes to affect the way a drug becomes absorbed, distributed, metabolized and excreted. These processes form the core principles of pharmacokinetics and explain why a drug behaves differently in the elderly as compared to the younger person. Proper consideration must be applied to every drug that is given to establish and ensure safety in an elderly patient.

Absorption

The gastrointestinal tract (GIT) changes with time and this greatly affects many orally administered drugs. Both the blood flow to and motility of the GIT reduce over time. The production of gastric acid reduces, allowing the pH to rise gradually. These affect gastric transit time and absorption of many drugs, which can be unpredictable in the elderly.

In the acute setting, many resuscitation drugs are given intravenously in the interest of speed and to avoid the problems associated with variability of absorption. If the subcutaneous and intramuscular routes are chosen, then reduced blood flow to the tissues may affect rate of absorption and onset of the drug given.

Distribution

The volume of distribution (V_d) in the body for many types of drugs changes as we grow older. One of the factors contributing to this is protein binding. Older adults produce less albumin, an important binding protein which determines the amount of free drug available for its actions, metabolism and clearance.

The aging process is also associated with lower total body water which results in a smaller V_d for water soluble drugs. The body muscle to fat ratio decreases as muscle mass declines over the years compared to the proportion of body fat. So, in general, drugs distributed in muscle may have a smaller V_d but vice versa for more lipid soluble drugs. Since V_d is an essential component of half life and loading dose, whenever V_d changes, so will both half life and loading dose. Sometimes the changes can be unpredictable. As an example, if V_d falls, the loading dose to achieve a desired drug concentration should also be reduced and the half life will also be altered. If these changes are not taken into consideration, then drug toxicity can easily occur. Many drugs lack precise information on how their V_d will be affected. Resuscitation drugs, sedatives and analgesics need to be given in lower doses and titrated to effect.

Metabolism

The metabolic capacity of the liver is reduced in the aging process. Generally, liver mass and intrinsic activity of hepatic enzymes such as CYP450 decline over the years resulting in delayed metabolism of many drugs and prolonging their half lives. Drugs are also presented to the liver for metabolism at a lower rate because of decreased hepatic blood flow. All of these normal physiological changes are made more pronounced by the effects of alcohol, poor nutrition and a regimen of multiple drugs with potential interactions and hepatotoxicity. Drugs need to be titrated to effect while actively monitoring for adverse effects, to maximise drug safety in the elderly.

Excretion

Most drugs are eliminated by the kidneys either unchanged or as metabolites. Similar to the liver, kidney function decreases with age because of reduction in the numbers of functioning glomeruli, renal mass and blood flow. However, unlike the hepatic system, the decline in renal function is more predictable and hence, the behaviour of drugs can be estimated by calculating the estimated glomerular filtration rate (see Chap. 15).

As a result of physiological changes in the elderly, the pharmacokinetic profiles of most drugs are altered significantly. Drugs have to be carefully considered and titrated according to each individual because accurate predictions of the changes can be difficult.

Pharmacodynamic Changes in the Elderly

The pharmacodynamics of many drugs can be altered substantially in the elderly. The response to drugs may be variable and the difference may be an exacerbated response or a diminished one. The mechanisms that may cause these effects are altered affinity of medications to the receptor sites, up regulation or down regulation of receptors that are readily available for drug-receptor activity. Some effects may be predictable. However, it is also not uncommon to observe unanticipated effects. The most clinically significant changes are seen in the central nervous system and cardiovascular system as compared to the other systems.

The arm-brain circulation time is usually prolonged in the elderly due to vascular changes and decreased cardiac output. Sedatives and analgesics may initially take a long time to produce and exhibit their effects but once their actions begin on their target effect sites, both wanted and unwanted actions will be pronounced. Excessive drowsiness, dizziness, confusion and even seizures have been frequently reported. These effects will cause additional morbidity in the elderly who are already ill to begin with on presentation. Meperidine is commonly used to manage pain in trauma cases and emergencies such as renal colic. Its metabolite normeperidine accumulates when renal clearance decreases. Therefore, if given without proper monitoring and titration, excessive drowsiness from the opioid effect as well as seizures due to the accumulated metabolite can result.

In the cardiovascular system, it is common for exaggerated hypotension, brady or tachyarrhythmias to occur. This increases the likelihood of adverse cerebrovascular and cardiovascular events occurring in this group of patients who are already at higher risk due to age or pre-existing medical conditions. Indirect acting vasopressors such as ephedrine may produce minimal or no effect when used to increase the mean arterial blood pressure because these individuals lack the neurotransmitters required for its action. Many long-standing diabetics also suffer from autonomic dysfunction and they do not respond well to resuscitation fluids or drugs. Caution must be given to the type and dose of many pharmacological interventions for acute care, in order to avoid unnecessary drug related mortality and morbidity.

Adverse Drug Reactions

Adverse drug reactions occur more frequently in the geriatric group of patients and the reactions are often more severe. For example, there will be greater risk of upper gastrointestinal bleeding, cardiovascular events and renal impairment in the elderly given non steroidal anti-inflammatory agents (NSAIDs) for pain. This could be due to the presence of a larger amount of unbound drug, from hypoalbuminaemia or pre-existing renal impairment decreasing clearance of the drug. In a report by the Centres for Disease Control and Prevention (CDC) in the United States, warfarin, insulin, anti-platelet drugs and oral hypoglycemic agents were listed, in that order, as the top most frequent causes of adverse drug reactions requiring emergency hospitalization.

Many older patients are on multiple medications for their existing medical problems and diseases. The interactions between these drugs commonly cause adverse drug reactions. It has also been reported that the risk of drug interaction rises with the number of medications taken by the patient. Dietary supplements can also interact with medications to cause adverse drug reactions, underlining the importance of taking a good history concerning both prescriptive and non prescriptive medications and substances. Some medications also aggravate the medical condition of pre-existing diseases in the elderly, such as diuretics exacerbating symptoms in patients suffering from incontinence.

Key Concepts

- There are few studies to provide true pharmacokinetic and pharmacodynamic effects of medication in the elderly.
- Most drug effects change significantly in the elderly.
- Administration of drugs to the elderly should be done slowly and titrated to effect.
- There are high risks of adverse drug reactions and drug interactions.
- The elderly patient should be closely monitored for responses to drugs.
- The elderly have little reserve so any morbidity is serious and can lead to mortality.

Summary

It is now increasingly common to provide acute care to the elderly in hospital because of a rapidly growing older population with many special needs. The changes in the pharmacokinetics and pharmacodynamics of drugs in the elderly are related to the physiological changes of aging. The multisystem degeneration in the body allows only a narrow margin of safety for many drugs. Drug administration and response is optimized by a thorough assessment of the patients' history, physiological and pathological status, and medication history. Every drug should be tailored to the individual's needs and there should also be a focus on anticipating and detecting undesired side effects.

Further Reading

1. Greenblatt DJ, von Moltke LL, Harmatz JS, et al. Pharmacokinetics, pharmacodynamics, and drug disposition. In: Davis KL, Charney D, Coyle JT, et al., editors. Neuropsychopharmacology: the fifth generation of progress. 5th ed. Philadelphia: Lippincott Williams & Wilkins; 2002. p. 507–24.
2. Handel J. Anaesthesia for the elderly. In: Allman KG, Wilson IH, editors. Oxford handbook of anaesthesia. 2nd ed. New York: Oxford University Press; 2006.
3. Hutchison LC, O'Brien CE. Changes in pharmacokinetics and pharmacodynamics in the elderly patient. J Pharm Prac. 2007;20:4–12.

4. Kirby BS, Crecelius AR, Voyles WF, Dinenno FA. Impaired skeletal muscle blood flow control with advancing age in humans: attenuated ATP release and local vasodilation during erythrocyte deoxygenation. Circ Res. 2012;111(2):220–30. doi:10.1161/CIRCRESAHA.112.269571.
5. Mangoni AA, Jackson SH. Age-related changes in pharmacokinetics and pharmacodynamics: basic principles and practical applications. Br J Clin Pharmacol. 2004;57:6–14.
6. Mendes-Nett RS, Silva CQ, Oliveira Filho AD, et al. Assessment of drug interactions in elderly patients of a family health care unit in Aracaju (Brazil): a pilot study. Afr J Pharm Pharmacol. 2011;5:812–18.
7. Miller SW. Therapeutic drug monitoring in the geriatric patient. In: Murphy JE, editor. Clinical pharmacokinetics. 4th ed. Bethesda: American Society of Health-System Pharmacists; 2007.
8. Routledge PA, O'Mahony MS. Woodhouse KW Adverse drug reactions in elderly patients. Br J Clin Pharmacol. 2004;57:121–6.
9. US Department of Health and Human Services. Elderly at risk of hospitalizations from key medications: promoting safe use of blood thinners and diabetes medications can protect patients. CDC Online Newsroom (US). 2011 Nov. Available from: www.cdc.gov/media/releases/2011/p1123_elderly_risk.html. Last accessed 15 June 2014.
10. Wooten JM. Pharmacology considerations in elderly adults. South Med J [Internet]. 2012;105 (8):437–45. Available from: http://www.medscape.com/viewarticle/769412_2. Last accessed 15 June 2014.

Chapter 22
The Obstetric Patient

Carolyn Chue-Wai Yim and Debra Si Mui Sim

Abstract Pregnancy causes a host of physiological changes which inevitably affect drug pharmacokinetics. Therefore, as drug absorption decreases while elimination increases during pregnancy, the plasma concentrations of drugs are generally reduced. Common drugs used in pregnancy include the uterotonics and uterine relaxants. Other drugs used are the antihypertensives and anticonvulsants (magnesium sulfate). As all drugs cross the placenta to some extent, and therefore some fetal exposure will occur, drug safety remains a concern in pregnancy due to the possible adverse effects of drugs, ingested by the mother, on the fetus. Drugs used during pregnancy have been classified according to how the drugs have been developed and their possible impact on the fetus. The ingestion of many drugs is considered 'safe' during breastfeeding because the concentrations of these drugs in the breast milk are usually low. When estimating risk during breastfeeding, factors to be taken into consideration include the infant dose, pharmacokinetics of the drug in the infant and effect of the drug in the infant.

Keywords Pregnancy • Pharmacokinetics • Medications used in pregnancy • FDA drug safety categorization • Lactation

Introduction

The pregnant patient is unique as there are always two lives to be considered when care is provided. Care providers must be especially cautious when administering medication to these patients, more so during the period of organogenesis in the first trimester, as the medication not only affects the patient but may also jeopardize the wellbeing of the fetus. This problem is further compounded by the fact that most

C.C.-W. Yim, M.B.B.S., M.Anaes (✉)
Department of Anesthesiology, Faculty of Medicine, University of Malaya,
50603 Kuala Lumpur, Malaysia
e-mail: drcarolyim@um.edu.my

D.S.M. Sim, B.Sc., Ph.D.
Department of Pharmacology, Faculty of Medicine, University of Malaya,
50603 Kuala Lumpur, Malaysia

© Springer International Publishing Switzerland 2015
Y.K. Chan et al. (eds.), *Pharmacological Basis of Acute Care*,
DOI 10.1007/978-3-319-10386-0_22

clinical drug trials have never conducted safety studies in this category of patients. Extrapolation from studies done on non-pregnant women (or from pre-clinical studies on pregnant animals) may not exactly reflect the responses likely to be found in pregnant patients, especially as there are marked physiological changes during the 9 months that are likely to impact on the pharmacology of the drugs administered.

Physiological Changes in Pregnancy

During pregnancy, physiological changes occur which involve all the systems. In the cardiovascular system, the cardiac output increases due to increased stroke volume and increased heart rate, while the total peripheral resistance decreases resulting in a decrease in the blood pressure. Progesterone results in a decrease in systemic vascular resistance by causing vasodilatation from early pregnancy. Plasma volume can increase up to 50 % at term resulting in a reduction of plasma protein concentration.

An increase in oxygen consumption leads to compensatory changes in the respiratory system. The tidal volume, respiratory rate and minute ventilation are increased. However, the functional residual capacity and residual volumes are decreased due to the enlarging uterus.

The glomerular filtration rate increases, hence increasing the clearance of urea, uric acid and creatinine. The parturient also has a higher risk of aspiration due to reduced lower esophageal sphincter tone and delayed gastric emptying.

Pharmacokinetic/Pharmacodynamic Changes in Pregnancy

The physiological changes of pregnancy may affect the pharmacokinetics and pharmacodynamics of drugs, and their resultant blood and tissue levels. Some of these changes may also have an impact on the fetus.

Absorption

High levels of progesterone result in delayed gastric emptying and reduced intestinal motility. This results in a longer time-to-peak plasma concentration (T_{max}) and there is an associated reduction in peak plasma concentrations (C_{max}). Bioavailability is only affected slightly. Although gastric pH is lowered in pregnancy, there is very little change in the absorption of most drugs ingested orally.

With regards to inhaled drugs, the increase in cardiac output and tidal volume results in an increased alveolar uptake of drugs, hence an increase in drug absorption through this route. Intramuscular absorption of drugs is also enhanced as a result of increased tissue perfusion.

Distribution

The expanded intravascular and extravascular water content also affect drug distribution. The volume of distribution of hydrophilic drugs is increased. The clinical effect is however minimal as there is also altered protein-binding.

The transfer of drugs across the placenta occurs mainly via passive diffusion with the rate-limiting step being placental blood flow. Large molecular weight (e.g., heparin), poor lipid solubility (e.g., gentamicin) and protein-bound drugs do not cross over to the fetus. However, those drugs that readily cross over into the fetal circulation may cause harmful effects to the developing fetus, especially during the first trimester. Examples of drugs that cause significant teratogenic or other adverse effects on the fetus include angiotensin converting enzyme inhibitors (ACE-I), warfarin, carbamazepine, phenytoin, tetracycline, iodide, heroin, tobacco, alcohol, thalidomide and diethylstilbestrol. To be considered teratogenic, a substance should (1) result in a characteristic set of malformations, (2) exert its effects at a particular stage of fetal development, and (3) show a dose-dependent incidence.

Recent studies have also demonstrated the presence of drug transporters in the placenta. These transporters pump a variety of drugs (e.g., anticancer drugs, viral protease inhibitors) back into the maternal circulation, thereby keeping the concentrations of these drugs low in the fetus. While keeping the toxic anticancer drugs away from the fetal circulation may be advantageous, preventing the transfer of viral protease inhibitors from the maternal circulation to the fetal circulation may however increase the risk of vertical transmission of HIV infection.

Metabolism/Elimination

Metabolism can be either increased or decreased depending on the drug concerned. For example, the metabolism of phenytoin is increased as both estrogen and progesterone induce the cytochrome P450 enzymes that metabolize phenytoin in the liver.

The renal blood flow is increased and this often results in an enhanced elimination of drugs. The slightly lower steady-state concentration of these drugs rarely requires an increase in dosage. For drugs with low therapeutic index, one has to be more vigilant and this may require blood level monitoring to guide appropriate dosing.

Drug Accumulation in the Fetus

As the metabolic enzyme activity of the liver in the fetus is always low and a large proportion of fetal blood bypasses the fetal liver, there are always concerns about drug accumulation in the fetus. Drugs which diffuse from the mother to the fetus also diffuse in the opposite direction depending on the concentration difference at the placenta. After delivery, the newborn has to rely on its immature kidneys and liver to eliminate drugs that have been administered just before delivery and care providers must be cognizant of this, especially while managing a parturient with a premature fetus.

Clinical Application: Ion Trapping

Labor epidurals allow the parturient to comfortably go through the process of labor. Local anesthetics, which are basic drugs, are commonly used with no detrimental effect on the unborn fetus. However, during instances when the fetus becomes acidotic, due mainly to hypoxia, more of these basic drugs become ionized. This ionized form becomes "trapped" and accumulates in the unborn fetus, a phenomenon known as "ion trapping". This situation further worsens the compromised fetus and can lead to an undesirable outcome.

Drugs Used During Pregnancy and Delivery

Care providers use many drugs for different purposes during pregnancy and delivery. There are some drugs that are used to aid with the delivery process and to manage some of the common conditions experienced during pregnancy.

Uterotonics

These are drugs used during parturition to stimulate uterine contractions to facilitate delivery, or after delivery to prevent uterine atony and postpartum hemorrhage.

Oxytocin

Oxytocin is secreted by the posterior pituitary gland. In pregnancy, it acts on the uterus resulting in uterine contraction. It also causes contraction of the milk ducts. In clinical practice, synthetic oxytocin can be given as an infusion during labor to augment uterine contractions, hence facilitating delivery. However hyperstimulation of the uterus can result in fetal compromise as the blood flow to the placenta will be affected by a persistently contracted uterus. Special precaution should be taken when oxytocin is used in parturients with cardiac disease as severe hypotension can occur. This is due to the vasodilatory effect of oxytocin.

In addition to stimulating/augmenting uterine contractions, oxytocin also has an anti-diuretic effect because of its structural similarity to vasopressin. Prolonged usage or high doses of oxytocin can cause potentially fatal water intoxication and hyponatremia. This is further exacerbated if it is infused in dextrose solutions.

When used for the induction of labor, an intravenous infusion is commenced and the dosage is increased at intervals of 30 min. The aim is to achieve 3–4 contractions every 10 min. The maximum rate is 0.02 units/min with not more than 5 units being administered over a 24-h period.

At delivery, oxytocin is used to prevent postpartum hemorrhage. A slow intravenous dose is given when the anterior shoulder is delivered or an intramuscular injection given after delivery of the fetus. An intravenous infusion after delivery can also be administered to treat postpartum hemorrhage.

Carbetocin

Carbetocin is a long-acting synthetic analogue of oxytocin. It can be administered intravenously as a single dose immediately following delivery, to prevent uterine atony and postpartum hemorrhage. Although less potent than oxytocin, carbetocin has a longer duration of action lasting up to about 1 h. Because it produces prolonged uterine contraction, carbetocin must not be administered before delivery of the infant. Its side effects are similar to those of oxytocin.

Ergometrine

Ergometrine is another drug which stimulates uterine contraction. A combination of oxytocin and ergometrine, known as Syntometrine, is given intramuscularly after delivery for the prevention of postpartum hemorrhage. Its side effects of severe nausea and vomiting make it a less favored drug. Furthermore, a marked rise in blood pressure can be observed due to vasoconstriction which can last for several hours. Hence, it can be hazardous if used in parturients with preeclampsia or cardiovascular diseases. Ergometrine 250–500 micrograms can be given by the intramuscular or intravenous route for the treatment of postpartum hemorrhage.

Drugs to Prevent Premature Uterine Contraction

Preterm birth is any birth that occurs before 37 weeks of gestation. Prematurity has a significant impact on both the survival and the long-term quality of life of the infant. Prevention and treatment of premature uterine contraction can improve the infant's outcome. Drugs that can be used include beta-agonists, calcium channel blockers, oxytocin receptor antagonists, prostaglandin synthetase inhibitors, and magnesium sulfate.

Salbutamol

Uterine relaxation is due to the β-adrenoceptor agonist effect. The drug binds to β-adrenoceptors which results in activation of the enzyme adenylate cyclase. This increases the levels of cAMP and causes a reduction of intracellular calcium level, resulting in uterine relaxation. Treatment for premature labor is via an intravenous infusion which is given in incremental doses at intervals of 10 min. The end point is when contractions diminish. The infusion rate is then further increased (maximum infusion of 45 μg/min) until contractions have ceased. This infusion rate is maintained for 1 h after which the infusion is gradually reduced. Side effects include tremor, hypotension, wheezing, tachycardia (at high doses), and hypoglycemia.

Calcium Channel Blockers

Nifedepine is given orally. The drug blocks the influx of calcium ions, resulting in uterine relaxation. Its side effects are headache, dizziness, constipation, and wheezing.

Prostaglandin Synthetase Inhibitors

These drugs act by depleting the amount of prostaglandins that initiate uterine stimulation. Examples of these drugs are indomethacin, aspirin and ibuprofen. Caution must be used with these drugs as they can result in premature closure of the ductus arteriosus in the fetus.

Oxytocin Receptor Antagonists

These drugs act by blocking oxytocin receptors in the myometrium. This prevents a rise in intracellular calcium which thus relaxes the myometrium. Adverse effects include nausea, vomiting, headache, chest pain and reaction at the injection site.

Drugs to Manage Preeclamptics

In managing the preeclamptic patient, many different drugs are used. Antihypertensive agents are used for the management of hypertension. Examples of such drugs are methyldopa, labetolol and nifedipine. Angiotensin-converting enzyme inhibitors (ACE-I) and angiotensin receptor blockers (ARB) are contraindicated in pregnancy due to fetal toxicity. In patients with severe preeclampsia, magnesium is used as first-line therapy for seizure prophylaxis.

Methyldopa

Methyldopa remains one of the most commonly used drugs for the treatment of hypertension in pregnancy. It is a prodrug which acts centrally as a α_2-adrenoceptor agonist. It is metabolized to α-methylnorepinephrine, which replaces norepinephrine in the neurosecretory vesicles of adrenergic nerve terminals. Its side effects include a decrease in mental alertness, impaired sleep and xerostomia.

Labetalol

This is a nonselective β-blocker and has the ability to also block vascular α_1-receptors. It can be given both orally and intravenously, making it extremely useful for the treatment of severe hypertension. Its side effects include fatigue, lethargy and peripheral vasoconstriction.

Magnesium Sulfate

Magnesium sulfate plays a major role in the treatment of eclampsia and its prevention. However special caution needs to be taken to prevent an overdose. Clinically, excessively high magnesium levels can present with the loss of patellar reflex, weakness, decrease of urinary output and hypotension. The antidote is calcium gluconate injection; which should be kept readily available.

An initial 4 gram (g) is given intravenously over a period of 5–15 min for both the prevention of seizure in preeclampsia and recurrent seizures in eclampsia. This is followed by continuing an intravenous infusion of 1 g/h for at least a period of 24 h.

Categorization of Drugs Used During Pregnancy According to Safety Concerns

An established categorization of drugs (Table 22.1) is made available by the Food and Drug Administration (FDA). These categories flag how the drugs have been developed and their likely impact on the fetus based on the research studies that have been done to determine their safe use in pregnancy and the unborn child. The categorization does not take into account the effects on breast milk.

A drug is considered approved for use in pregnancy only if the words "pregnancy", "obstetrics", "labor", "delivery" or "lactation" appears in the "Indications" section of the drug packaging. Furthermore, failure to mention that the drug is contraindicated for use in pregnancy does not imply that it is safe for use. Currently, ritodrine is the only drug that is specifically approved for use in pregnancy. It is also acknowledged that potential adverse effects with regards to the development of the

Table 22.1 Categorization of drugs used during pregnancy according to the Food and Drug Administration (FDA)

Category A	Adequate and well-controlled studies have failed to demonstrate a risk to the fetus in the first trimester of pregnancy (and there is no evidence of risk in later trimesters).
Category B	Animal reproduction studies have failed to demonstrate a risk to the fetus and there are no adequate and well-controlled studies in pregnant women.
Category C	Animal reproduction studies have shown an adverse effect on the fetus and there are no adequate and well-controlled studies in humans, but potential benefits may warrant use of the drug in pregnant women despite potential risks.
Category D	There is positive evidence of human fetal risk based on adverse reaction data from investigational or marketing experience or studies in humans, but potential benefits may warrant use of the drug in pregnant women despite potential risks.
Category X	Studies in animals or humans have demonstrated fetal abnormalities and/or there is positive evidence of human fetal risk based on adverse reaction data from investigational or marketing experience, and the risks involved in use of the drug in pregnant women clearly outweigh potential benefits.
Category N	FDA has not classified the drug.

central nervous system can occur if drugs are trapped in the brain of the fetus. There is no requirement for drug manufacturers to conduct follow-up studies to determine the long-term outcome on the exposed fetus. When prescribing medications for the pregnant patient, it is important to be cognizant of the category of safety concerns the drug has been subjected to during its development.

Clinical correlation – The Thalidomide story

Drug development is a long and costly process, but it is important that a new drug undergoes rigorous preclinical testing and phase I, II and III clinical trials before it can be registered for use.

Thalidomide was first sold in West Germany in 1957 as a sedative or hypnotic. It was also used to relieve nausea and morning sickness in pregnant women. It was available as an over the counter drug in Germany in the late fifties. This easy access to the drug by pregnant mothers and failure to determine the impact of the drug during pregnancy resulted in approximately 5,000 to 7,000 babies born with malformation of the limbs (phocomelia), of which only 40% survived. This unfortunate incident led to the development of tighter drug regulations concerning drug development and drug use worldwide.

Use of Drugs During Lactation

Most drugs administered to lactating mothers can be detected in the breast milk and this may carry a risk to a breastfed infant. However, the concentrations of these drugs in the breast milk are usually low. Exposure of infants to drugs during breast-feeding is almost always less than exposure during pregnancy. Therefore, the ingestion of many drugs is considered 'safe' during breastfeeding. However, the ingestion of even small amounts of some highly toxic drugs (e.g., cytotoxic agents, radioactive iodine, immunosuppressive agents) by lactating mothers should be best avoided, in view of the potential risk these drugs carry on the infant.

Most drugs enter breast milk by passive diffusion of the free (unbound) and unionized form. The transfer of drugs into breast milk is therefore greatest in the presence of low maternal plasma protein binding and high lipid solubility. As breast milk is slightly more acidic (pH 7.2) than plasma (pH 7.4), weakly basic drugs transfer more readily into breast milk and become trapped secondary to ionization. Varying fat content within and between feeds may also affect transfer of drugs into breast milk. Thus, when estimating risk during breastfeeding, factors to be taken into consideration include the infant dose, pharmacokinetics of the drug in the infant and effect of the drug in the infant.

To minimize potential adverse effects caused by drugs taken by lactating mothers, it is advisable to take the medication 30–60 min after nursing and 3–4 h before the next feeding. This allows time for many drugs to be cleared from the nursing mother's blood and the infant plasma concentration of the drug will be relatively low.

Key Concepts

- The pharmacokinetics of drugs is affected by pregnancy as a result of the physiological changes that occur during pregnancy.
- Drugs used during pregnancy and delivery are divided into uterine stimulants, relaxants and drugs used to treat or prevent preeclampsia/eclampsia.
- Use of drugs in pregnancy can be guided by the Food and Drug Administration (FDA) categorization of drugs with safety recommendations.
- When estimating risk during breastfeeding, factors to be taken into consideration include the infant dose, pharmacokinetics of the drug in the infant and effect of the drug in the infant.

Summary

Treating the parturient remains a challenge for the clinician as it involves treating two lives. This dilemma is compounded by the fact that there is a lack of clinical drug trials that study the safety of drugs used during pregnancy. The physiological changes experienced by the pregnant woman results in changes in drug pharmacokinetics. Drug absorption, distribution, metabolism and elimination may be affected. The risk of accumulation of basic drugs in the fetus is possible especially in premature acidotic babies, in the phenomenon of "ion trapping". Drugs used during pregnancy and delivery include the uteronics, uterine relaxants and drugs for the treatment and prevention of preeclampsia/eclampsia. All drugs cross the placenta to some extent, and therefore some fetal exposure will occur. The risk/benefit ratio must be considered when drugs are used in pregnancy, especially during the first trimester. The ingestion of many drugs is considered 'safe' during breastfeeding because the concentrations of these drugs in the breast milk are usually low.

Further Reading

1. Dawes M, Chowienczky PJ. Pharmacokinetics in pregnancy. Best Pract Res Clin Obstet Gynecol. 2000;5(6):819–26.
2. Gardiner S. Drug safety in lactation. Prescriber Update 2001 May;21:10–23.
3. Haire DB. How Obstetric drugs can affect labor and birth [Internet]. 2000. Available from http://www.aimsusa.org/rothdrug.htm. Last accessed 8th Sept 2014.
4. Haire DB. FDA approved obstetrics drugs: their effects on the mother and baby [Internet]. 2001. Available from http://www.aimsusa.org/obstetricdrugs.htm. Last accessed 8th Sept 2014.
5. Podymow T, August P. Update on the use of antihypertensive drugs in pregnancy. Hypertension. 2008;51:960–9.
6. Stoelting RK, Hillier SC, editors. Pharmacology & physiology in anaesthetic practice. 4th ed. Philadelphia: Lippincott Williams & Wilkins; 2006.
7. Weinberg GL. Lipid emulsion infusion. Anesthesiology. 2012;117:180–7.

Chapter 23
The Obese Patient

Hou Yee Lai

Abstract Obese patients have a different body composition compared to a normal weight individual. With increasing obesity, fat mass will increase at a greater rate than lean body weight. As a result, an obese patient will have greater fat mass but only slightly more muscle mass than a normal weight individual of the same height and sex. Due to these differences, drug kinetics and dynamics are altered in the obese. Drug administration practices need to be altered to accommodate recommended doses based on studies done in normal weight subjects. These adjustments are done using different weight descriptors such as total body weight (TBW), lean body mass (LBM) or ideal body weight (IBW). Choice of weight descriptor to guide dosing depends on the lipophilic or hydrophilic property of the drugs. Generally, hydrophilic drugs will primarily redistribute to the muscle compartment, thus the loading dose of these drugs should be based on LBM. Likewise, lipophilic drugs will eventually redistribute to the fat mass, thus loading dose of these drugs should be based on TBW. However, other factors such as protein binding, tissue binding and degree of ionization also greatly influence how drugs distribute in the different tissue compartments. As a result of this complex interaction, drug dosing should ultimately be based on the available evidence for that particular drug. Factors such as metabolism, clearance and pharmacodynamic idiosyncrasies must be taken into consideration for any individual. The titration of a drug aided by drug level monitoring or monitoring of response to the drug may be good practice. This is especially so when using drugs with a narrow therapeutic window, and in obese patients with possible life threatening complications of obesity such as obstructive sleep apnea.

Keywords Obesity • Body composition • Weight descriptors • Total body weight • Lean body mass • Ideal body weight • Lipophilic drugs • Hydrophilic drugs • Drug titration • Narrow therapeutic window

H.Y. Lai, M.B.B.S., M.Anaes (✉)
Department of Anesthesiology, Faculty of Medicine, University of Malaya,
50603 Kuala Lumpur, Malaysia
e-mail: houyee@um.edu.my

© Springer International Publishing Switzerland 2015
Y.K. Chan et al. (eds.), *Pharmacological Basis of Acute Care*,
DOI 10.1007/978-3-319-10386-0_23

191

Introduction

Obesity is becoming more prevalent globally. As the number of obese patients increases in our practice, it is necessary to know how to deliver care according to their needs. These needs may be different from those of a patient of normal weight and particular attention needs to be paid to their medication.

The pharmacokinetics and possibly pharmacodynamics of drugs are altered from the norm. There are few studies to directly show how these changes occur as the obese population is excluded from many drug development trials. How we administer medication is guided by studies done on normal weight patients. We have to adjust the needs of the obese patients using different weight descriptors.

Obesity and Body Composition

Obesity is defined by the World Health Organization (WHO) as a body mass index (BMI) of greater than 30 kg/m^2 (Table 23.1).

Body composition of the obese is vastly different from their normal weight counterparts (Fig. 23.1). With increasing obesity, fat mass will increase at a greater rate than lean body weight. In other words, an obese person will have a greater fat mass, but only slightly more muscle mass than a normal weight individual of the same height and sex.

Table 23.1 Classification of the grades of obesity, by the body mass index

Classification	BMI (kg/m^2)
Underweight	<18.5
Normal weight	18.5–24.99
Overweight	25.00–29.99
Class 1 obesity	30.00–34.99
Class 2 obesity	35.00–39.99
Class 3 obesity	40.00–44.99

Adapted from: Preventing and managing the global epidemic of obesity. Report of the WHO consultation of obesity. WHO, Geneva June 1997

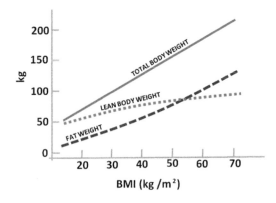

Fig. 23.1 Relationship between total body weight, lean body weight and fat weight to body mass index, in a standard height male (With permission from Oxford Publishers, BJA 2010;105(S1):i16–i23)

Total body weight (TBW) is the sum of lean body mass (LBM) and fat mass. Most drug doses recommended by the manufacturer are based on TBW.

LBM is accurately determined using dual energy x-ray absorptiometry (DEXA) but is more practically estimated from TBW by using the following formulas:-

$$\text{For males} \quad (1.1 \times \text{TBW}) - 128(\text{TBW}/\text{Height})^2$$

$$\text{For females} \quad (1.07 \times \text{TBW}) - 148(\text{TBW}/\text{Height})^2$$

Ideal body weight (IBW) was introduced by Devine in 1974 to better predict the clearance of certain drugs in the obese. He estimated IBW according to the following formulas:-

$$\text{For males} \quad 50 + 2.3(\text{Height [Inches]} - 60)$$

$$\text{For females} \quad 45.5 + 2.3(\text{Height [Inches]} - 60)$$

Administration of certain drugs based on TBW in the obese may result in toxicity. On the other hand, dosing based on LBM or IBW may result in under dosing for some drugs. Understanding how the drugs distribute in the body and using the correct weight descriptor will be the key to safely adapting drug doses in this population.

Body Composition and the Impact on Pharmacokinetics

Body composition, although changed in the obese, affects only certain pharmacokinetic parameters. This is mainly in the way drugs are distributed.

Absorption

Oral drug absorption is not altered by obesity. Lipophilic drugs may take longer to reach steady state than hydrophilic ones as there is increased fat mass, taking these drugs longer to saturate the fat compartment.

Subcutaneous and intramuscular routes of administration may have an unpredictable effect in the obese. There may be differences in blood flow to the skin and drugs intended to be intramuscular may be deposited subcutaneously due to insufficient needle length.

Distribution of Drugs

The difference in body composition will influence distribution of many drugs at equilibrium. Initial distribution of intravenous drugs will not be much different than a normal weight individual as this depends on the cardiac output. All intravenous drugs will be distributed rapidly to the major organs and muscle mass.

After redistribution, lipophilic drugs tend to accumulate in the fat mass. As a result, volume of distribution (V_d) for lipophilic drugs will increase in the obese. The loading dose of these drugs should therefore be based on TBW.

Hydrophilic drugs will be primarily distributed to the muscle mass. As there is only modest increase in LBM with increasing TBW, the V_d of hydrophilic drugs will only increase slightly in the obese and hence LBM should be used to guide dosing of these drugs.

Despite this understanding, other factors such as protein binding, tissue binding and ionization of drugs may influence how drugs distribute to the different tissue compartments of the obese patient. It is best to use this as a guide and refer to the available literature of the individual drug for dosage guidance.

Metabolism and Clearance of Drugs

Clearance of drugs is controlled by the function of the liver and kidney.

Cytochrome P450 2E1 activity and phase 2 conjugation activities in the liver may be increased in the obese. However, fatty liver, which is a common condition in the obese, may counter this increase. It is still unclear how this may alter clinical practice.

There is limited knowledge regarding the effects of obesity on kidney function. In the obese patient, it is better to determine the actual individual creatinine clearance in order to determine the maintenance dose of a drug.

Clearance half life of lipophilic drugs will usually increase as a result of increased V_d. However, for hydrophilic drugs, clearance half life in the obese may be no different from the normal weight individual. Again, these are just rules of thumb, and certain drugs may prove to be exceptions.

Clinical Correlation – Prescribing the Correct Dose

Administration of acetaminophen in the obese child for fever may prove dangerous if it is based on total body weight. The recommendation of 15–20 mg/kg 4–6 hourly up to a limit of 60 mg/kg/day is based on studies of normal weight children. If an 8 year old boy who weighs 50 kgs is to be prescribed acetaminophen based on his total body weight, his prescription would be 750 mg 6 hourly. A typical 8 year old boy should weigh around 25 kgs according to a typical height and weight chart for boys.

This prescription would put the 50 kg child in danger of acetaminophen overdose and liver toxicity as the liver metabolic capacity may be overwhelmed. As there is currently no guideline to determine the correct dose of acetaminophen for the obese child, it may be prudent to prescribe based on ideal body weight.

A concomitant prescription of ibuprofen (again based on ideal body weight) if there is no contraindication may bring about a better antipyretic effect than acetaminophen alone.

Pharmacodynamic Changes of Drugs in the Obese

If the knowledge of the pharmacokinetics of drugs is limited in the obese, the knowledge of pharmacodynamics of drugs is even more limited.

Drugs which have a narrow therapeutic window should be used with caution. Drugs which depress the conscious level of the patient in particular should be titrated according to needs. Depression of conscious level can bring about life threatening obstruction of the airway and associated hypoventilation and hypoxia.

Similarly, drugs that depress breathing e.g., narcotics should be used cautiously. Patients who receive such drugs must be closely monitored for worsening of sleep apnea which is commonly associated with obesity, and the resultant reduction in oxygen delivery.

Dosing Guides for the Obese

Database on dosing for obese patients is limited because of their exclusion from most clinical trials. Some drug trials on the obese have attempted to adjust drug dosing based on weight descriptors and other descriptors like BMI and body surface area (BSA). To date, various weight descriptors such as TBW, LBM, IBW, adjusted body weight (ABW), percent of IBW (%IBW) and predicted normal weight (PNWT) have been studied with varying degrees of success.

No single descriptor can be applied successfully across the board for all pharmacological agents. Some studies generate conflicting results and it is currently recommended to adjust dosing based on the individual drug and available evidence.

Titration

Titration may be the best measure especially for those drugs with a narrow therapeutic window. This is best aided by the use of blood level monitoring of the drug to guide further dosing, or the use of response to the drug as a means of getting as close as possible to meeting the needs of these patients.

Examples of this include checking of international normalised ratio (INR) after starting warfarin, assay of drug levels after administering antibiotics such as vancomycin, monitoring of muscle twitch height using Train of Four (TOF) count after administering muscle relaxants and using bispectral index (BIS) monitors to monitor depth of anesthesia with propofol infusions.

Key Concepts

- Pharmacokinetics of drugs are altered in the obese due to large increase in fat mass and a more modest increase in lean body mass with increasing total body weight.
- The initial distribution of drugs following intravenous administration is no different from a normal weight individual as it follows the distribution of the cardiac output.
- After redistribution, the volume of distribution of lipophilic drugs will greatly increase, thus increasing the clearance half life.
- The volume of distribution of hydrophilic drugs is similar to that found in the normal weight individual.
- The clearance of drugs is dependent on renal and liver function, which may be enhanced or impaired by associated comorbidity in the obese patient.
- There is no single descriptor that is best applied to adjust the dose of various drugs and an individualized approach is recommended with titration of drugs being a good option.

Summary

Despite understanding the principles outlining the distribution and elimination of different drugs in the obese, there is considerable variation in drug pharmacokinetics and pharmacodynamics. Until more evidence emerges to guide drug dosing in the obese with the use of the different weight descriptors, more care must be taken particularly when using drugs with a narrow therapeutic window. Titration of drugs with monitoring of response may be the best option to guide therapy.

Further Reading

1. Devine BJ. Gentamicin therapy. Drug Intell Clin Pharm. 1974;8:650–5.
2. Green B, Duffull SB. What is the best size descriptor to use for pharmacokinetic studies in the obese? Br J Clin Pharmacol. 2004;58(2):119–33.
3. Hanley M, Abernethy DR, Greentblatt DJ. Effect of obesity on the pharmacokinetics of drugs in humans. Clin Pharmacokinet. 2010;49(2):71–87.
4. Ingrande J, Lemmens HJM. Dose adjustment of anaesthetics in the morbidly obese. Br J Anaesth. 2010;105(S1):i16–23. doi:10.1093/bja/aeq312.
5. Pai M. Drug dosing based on weight and body surface area: mathematical assumptions and limitations in obese adults. Pharmacotherapy. 2012;32(9):856–68.
6. World Health Organization. Obesity: preventing and managing the global epidemic. Report of a WHO consultation of obesity. Geneva, 3–5 June 1997. Also available from: http://whqlibdoc.who.int/hq/1998/WHO_NUT_NCD_98.1_%28p1-158%29.pdf. Last accessed 16 June 2014.

Chapter 24
The Bleeding Patient

Ina Ismiarti Shariffuddin

Abstract In the bleeding patient, the first line of management is to replace the blood loss with fluids or blood/blood products in a timely manner to ensure that perfusion to the vital organs is maintained. Securing hemostasis through surgical or other interventions at the earliest opportunity is important to enable successful volume replacement. Coagulopathy associated with inadequate replacement will worsen the bleeding. Unless torrential bleeding is arrested, efforts to replace blood volume and oxygen carrying capacity of blood will fail. The type of coagulopathy can be determined by the use of thromboelastography or plasma based coagulation tests in the laboratory. This will allow appropriate selective component replacement therapy. In those with bleeding as a result of trauma, balanced blood transfusion is advocated. Drugs that improve coagulation work at either the coagulation pathway or at the fibrinolytic pathway. Tranexamic acid prevents the activation of plasminogen to plasmin. Aprotinin inhibits the proteolytic enzymes and prevents fibrinolysis. Activated Factor Vlla increases clot formation at sites of exposed tissue factor.

Keywords Blood products • Thromboelastography • Hemostasis • Coagulation • Tranexamic acid • Activated Factor Vlla • Aprotinin

Introduction

Patients who are bleeding can present with massive life threatening hemorrhage or slow, continuous ooze. In emergency trauma, bleeding has been shown to be a leading cause of death. In order to treat a bleeding patient in a timely and cost effective manner, it is important for the physician to obtain a focused history, perform a complete physical examination and order appropriate laboratory tests.

Whilst blood and blood products are important for managing these patients, there are also drugs that have been shown to be useful in reducing blood loss.

I.I. Shariffuddin, MBChB, M.Anaes (✉)
Department of Anesthesiology, Faculty of Medicine, University of Malaya,
50603 Kuala Lumpur, Malaysia
e-mail: ismiarti@ummc.edu.my

© Springer International Publishing Switzerland 2015
Y.K. Chan et al. (eds.), *Pharmacological Basis of Acute Care*,
DOI 10.1007/978-3-319-10386-0_24

These drugs play various roles in the coagulation and thrombotic processes and when used appropriately can significantly reduce the need for blood products.

Pathophysiology

In the state of health, the body has a hemostatic mechanism whereby minor injury will not result in excessive blood loss. Primary hemostasis is the formation of a platelet plug at the site of injury, which occurs within seconds of the injury. Secondary hemostasis, that takes a few minutes to complete, involves the coagulation system and results in fibrin formation leading to the arrest of the bleeding. This process is illustrated in Fig. 24.1. Abnormal bleeding can be caused by abnormal platelet function, disorders of clotting factors or blood vessels and fibrinolytic defects (Table 24.1).

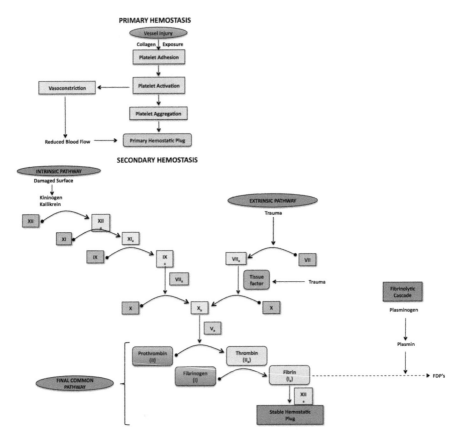

Fig. 24.1 Primary hemostasis, secondary hemostasis and the fibrinolytic pathway. FDP - fibrin degradation products

Table 24.1 Causes of bleeding disorders

Disorder of platelet function	**Inherited** Glanzmann thrombasthenia Von Willebrand's disease **Acquired** Drug-induced eg: anti platelet drugs, NSAIDs Liver disease Alcoholism
Disorders of clotting factors	**Inherited** Hemophilia A and B Von Willebrand's disease Factor VII deficiency Factor V deficiency Factor II deficiency Factor X deficiency Factor XI deficiency Factor XIII deficiency **Acquired** Liver disease Renal disease Vitamin K deficiency Drug induced eg: heparin, warfarin
Fibrinolytic defects	Disseminated intravascular coagulation
Disorder of blood vessels	Scurvy Purpura Hereditary hemorrhagic telangiectasia Cushing syndrome Ehlers-Danlos syndrome

Management of Haemorrhage

In treating the acutely bleeding patient, management should focus on early recognition of blood loss, rapid control of the source of bleeding and restoration of circulating blood volume. Emergency intervention to stop the source of blood loss as soon as possible (e.g. surgery) is of paramount importance for a good outcome. Meanwhile, blood, blood products, drugs and volume replacement can be administered as required to support the circulation.

Adequate blood pressure, circulatory volume and hemoglobin are necessary to enable oxygen to be delivered to the tissues. Prolonged hypotension or shock resulting from blood loss can become irreversible, due to tissue hypoxia and acidosis. Besides management of volume and oxygen delivery, the hemostatic component of blood has to be corrected as well to maintain the circulation, once

Table 24.2 Blood components that can be used to restore the circulation

Products	Indications	Dose
Whole blood	Used for red cell replacement in acute blood loss with hypovolemia	1 unit of whole blood increases hemoglobin by 0.75 –1 gm/dl and haematocrit by 3–5 %.
Packed cells	Replacement of red cells in normovolemic anaemia, acute or chronic for restoration of oxygen carrying capacity. It increases oxygen carrying capacity without increasing blood volume.	1 unit of packed cell increases hemoglobin by 0.75 – 1 gm% and hematocrit by 2.5 – 3 %.
Platelet concentrate	Treatment of bleeding due to thrombocytopenia and platelet function defects. Prevention of bleeding due to thrombocytopenia.	5 – 10 ml/kg will raise platelet count by 50 – 100 x 109 /L
Fresh frozen plasma	Used in patients with multiple coagulation factor deficiencies 1. Liver diseases 2. Warfarin (anticoagulant) overdose 3. Depletion of coagulation factors in patients receiving large volume transfusions 4. Disseminated Intravascular Coagulation (DIC) 5. Thrombotic Thrombocytopenic Purpura (TTP)	10–20 ml/kg of body weight
Cryoprecipitate	Use as an alternative to factor VIII concentrate in the treatment of inherited deficiencies of: 1. Von Wille brand factor (Von Wille brand disease) 2. Factor VIII (hemophilia A) 3. Factor XIII 4. As a source of fibrinogen in acquired coagulopathies e.g. Disseminated Intravascular Coagulation (DIC)	5–10 ml/kg of bodyweight.

the source of bleeding has been stopped. The blood components that can be used to restore the coagulation profile, oxygen carrying capacity and circulatory volume are as listed in Table 24.2.

Colloids and crystalloids can be used to restore the circulation, provided the level of hemoglobin in the patient's circulation is sufficient for oxygen delivery. These colloids and crystalloids are discussed in Chap. 29. Blood products such as platelets, fresh frozen plasma (FFP) and cryoprecipitate should not be used as volume replacement. Instead their use should be targeted to replace the deficient components. The decision as to which blood component needs to be administered can be guided by point of care testing such as Thromboelastography (TEG) or plasma based coagulation tests (PBCT).

Monitoring of Coagulation Disorders

Thromboelastography Versus Plasma Based Coagulation Tests

Thromboelastography (TEG) has been shown to provide faster results in comparison to the plasma based coagulation tests such as the platelet count, prothrombin time, fibrinogen, antithrombin and D-dimer. TEG provides rapid information about hemostatic changes in the presence of significant blood loss and can provide valuable indications for specific blood product therapy. This is advantageous in the setting of on-going hemorrhage. However, laboratory analysis of plasma based coagulation tests such as fibrinogen and antithrombin provide better information about the specific factors that cause impaired hemostasis. Fibrinogen has been shown to be the most important factor in controlling bleeding and is the first factor to decrease to critical levels in massive bleeding.

Hemostatic Resuscitation

Recently, the concept of hemostatic resuscitation has been introduced. Early pre-emptive balanced transfusion therapy is advocated. The blood and blood products are recommended to be given in a fixed red blood cells (RBC): FFP: Platelet ratio of 1:1:1 in the early stage of massive bleeding. As bleeding is a dynamic process, the transfusion therapy can then be goal directed according to the point of care test of TEG or the PBCTs.

Hemostatic Drugs

Hemostatic drugs are agents that have been shown to reduce the need for blood products. The various products work at different sites in the coagulation pathway and the fibrinolytic pathway.

Drugs Inhibiting Fibrinolysis

(i) Tranexamic Acid

This lysine-like drug reversibly blocks the binding sites of plasminogen, thus preventing its activation to plasmin. In trauma, 1 g can be given intravenously within 3 h of the event followed by 1 g infused over 8 h. For the treatment of postpartum haemorrhage, 1 g can be given intravenously followed by a further 1 g if bleeding continues or recurs. In the international CRASH-2 trial (randomised trial of more than 20,000 patients with major traumatic haemorrhage), it was shown that the administration of Tranexamic acid within 3 h of trauma caused a significant reduction in mortality in patients without any increase in thromboembolic events.

(ii) Aprotinin

Aprotinin use is aimed at reducing the need for blood transfusion during surgery which involves high risk of significant blood loss. It inhibits many proteolytic enzymes thus reducing fibrinolysis. A loading dose of 1–2 million KIU is administered as a slow intravenous injection or infusion over 20–30 min after induction of anesthesia. The use of Aprotinin is now limited to a small population of patients with a particularly high risk of bleeding because of safety issues, such as anaphylactic reactions (1 in 200 patients). Studies show greater risk of mortality and morbidity with the use of aprotinin compared to other anti-fibrinolytics.

Drugs Acting on the Coagulation Cascade

(i) Recombinant Factor V11A

This directly activates blood-clot formation at sites of exposed tissue factor in damaged blood vessels. The use of recombinant factor VII A (rFVIIa) to reduce blood transfusion is controversial. It does not reduce mortality but only causes a modest reduction in blood loss or transfusion when used prophylactically or therapeutically in patients without hemophilia [6]. It was also shown that higher doses were no more effective and were not a substitute for coagulation factor replacement. In fact, the results indicate that the incidence of arterial thrombosis is high in patients receiving rFVIIa. Therefore, the routine use of rFVIIa for non-haemophilia bleeding cannot be recommended outside well-designed clinical trials.

(ii) Desmopressin

Desmopressin (DDAVP) is mainly used to treat or prevent bleeding in patients with mild type I von Willebrand's disease or hemophilia A. It causes the release of Factor VIIIc and von Willebrand factor (vWF) from endothelial cells. Desmopressin may also reduce bleeding in patients with uraemia and platelet dysfunction due to kidney failure. It causes transient increase in factor VIII level through the release of endogenous factor VIII in patients with hemophilia A and von Willebrand's disease. The dose is 0.3 mcg/kg given intravenously or subcutaneously.

Key Concepts

- Timely transfusion of fluids or blood/blood products in a bleeding patient aided by damage control surgery, are essential in the successful management of massive blood loss.
- Blood or blood product therapy can be facilitated by appropriate point of care monitoring using thromboelastography or laboratory diagnosis using plasma based coagulation tests.
- Hemostatic drugs work on the coagulation pathways to assist coagulation or the fibrinolytic pathways to slow the lysis of the definitive clot.

Summary

In the bleeding patient, the first line of management is the timely replacement of blood volume with fluids or blood/blood products. Damage control surgery in the trauma patients is also essential to arrest bleeding to enable restoration of the circulating blood volume and reduce life threat. The type of coagulopathy can be determined with the use of thromboelastography or plasma based coagulation tests in the laboratory. This will enable a correct diagnosis of coagulation factor deficiencies and allow appropriate selective component therapy. In those with bleeding as a result of trauma, balanced blood transfusion is advocated. Drugs to improve hemostasis work at either the coagulation pathway or at the fibrinolytic pathway. Tranexamic acid prevents the activation of plasminogen to plasmin. Aprotinin inhibits the proteolytic enzymes and prevents fibrinolysis. Activated Factor VIIa increases clot formation at sites of exposed tissue factor.

Further Reading

1. CRASH-2 Collaborators, Roberts I, Shakur H, Afolabi A, Brohi K, Coats T, Dewan Y, et al. The importance of early treatment with tranexamic acid in bleeding trauma patients: an exploratory analysis of the CRASH-2 randomised controlled trial. Lancet 2011;377(9771):1096–101, 1101. e1-2. http://dx.doi.org/10.1016/S0140-6736(11)60278-X. Last accessed on 9 Sept 2014.
2. Gupta A, Epstein JB, Cabay RJ. Bleeding disorders of importance in dental care and related patient management. J Can Dent Assoc. 2007;73(1):77–83.
3. Joint United Kingdom (UK) Blood Transfusion and Tissue Transplantation Services Professional Advisory Committee. Available at http://www.transfusionguidelines.org.uk. Last accessed on 12 June 2014.
4. Karlsson O, Jeppsson A, Hellgren M. Major obstetric haemorrhage: monitoring with thromboelastography, laboratory analyses or both? Int J Obstet Anesth. 2014;23:10–7.
5. Rourke C, Curry N, Khan S, et al. Fibrinogen levels during trauma hemorrhage, response to replacement therapy, and association with patient outcomes. J Thromb Haemost. 2012;10:1342–51.

6. Simpson E, Lin Y, Stanworth S, Birchall J, Doree C, Hyde C. Recombinant factor VIIa for the prevention and treatment of bleeding in patients without haemophilia. Cochrane Database Syst Rev 2012, Issue 3. Art. No.: CD005011. http://dx.doi.org/10.1002/14651858.CD005011.pub4. Last accessed on 9 Sept 2014.
7. Stensballe J, et al. Viscoelastic guidance of resuscitation. Curr Opin Anesthesiol. 2014;27(2):212–8.

Chapter 25
The Septic Patient

Suresh Kumar

Abstract Improving the outcome of critically ill septic patients requires giving appropriate antibiotic(s), as early as possible and at correct doses. Empirical antibiotic therapy will require knowledge of possible anatomic site of infection, local resistance patterns and presence of any of the risk factors for multi-resistant pathogens. The more severe the infection, the more urgent is the need to administer antibiotics. Combination antibiotic therapy will sometimes be required especially if the patient has risk factors for multi-resistant pathogens. After 48–72 h of treatment, earnest attempt has to be made to de-escalate especially when the sources of infection with sensitivities are known. Duration of antimicrobial therapy often has to be tailored according to the treatment response. Optimising the dose of antibiotic therapy requires knowledge of the alterations to their pharmacokinetics and pharmacodynamics in the critically ill setting. Volume of distribution and clearance are especially affected for the hydrophilic antibiotics. This entails the need to administer loading doses for some of the concentration dependent and time dependent antibiotics and for more frequent dosing for time dependent antibiotics. Hypoalbuminemia which is common in the critically ill patients, results in the alteration in the pharmacodynamics of highly protein bound antibiotics. In such situations, highly protein bound antibiotics will need to be given in higher dose or more frequently.

Keywords Appropriate antibiotic • Combination antibiotic therapy • Volume of distribution • Hypoalbuminemia • Hydrophilic antibiotics • Lipophilic antibiotics • Pharmacokinetics • De-escalation of antibiotic • Empirical antibiotic therapy

Introduction

The septic patient features prominently in acute care medicine. Choosing an appropriate antibiotic is crucial to prevent mortality. This involves giving an initial antimicrobial that is appropriate against the bacteria concerned, as early as possible

S. Kumar, M.B.B.S., MRCP, PG Diploma Epid (✉)
Hospital Sungai Buloh, Jalan Hospital, Sungai Buloh 47000, Selangor, Malaysia
e-mail: chikku.suresh@gmail.com

© Springer International Publishing Switzerland 2015   205
Y.K. Chan et al. (eds.), *Pharmacological Basis of Acute Care*,
DOI 10.1007/978-3-319-10386-0_25

and in doses that ensure therapeutic levels are achieved in the blood and at the target site even with the first dose. In intensive care settings early and adequate antibiotic therapy has significant mortality benefits. Causal relationship exists too between inappropriate dosing, clinical outcome and development of resistance. There is significant interrelationship between pathophysiological changes in septic patients and how these may affect the pharmacokinetics and pharmacodynamics of administered antibiotics.

Appropriate Empirical Antibiotic Therapy

Often in the septic patient in critical care, antibiotics need to be given even before the specific pathogen causing the infection is known. Inappropriate antibiotic therapy is defined as administration of antimicrobial to which the pathogen responsible for the infection is resistant. Inappropriate therapy is associated with increased mortality in critically ill patients. In addition, delay in the initiation of appropriate antimicrobials is associated with increased mortality. The sicker the patient, the more urgent is the need to administer effective antimicrobial therapy (Table 25.1). Selection of appropriate antimicrobial therapy requires knowledge of possible anatomic site of infection, risk factors for infection with multi-resistant pathogens, and local microbiological flora and resistance patterns. Risk factors for multi-resistant pathogens include prolonged or prior hospital stay, prior colonisation or infection with multi-resistant pathogens, and recent antimicrobial use.

Local and unit-specific antimicrobial guidelines need to be constructed after considering the above mentioned factors. Use of such guidelines increases the chance of achieving adequate empiric coverage. Such guidelines have to be updated regularly after considering changes in the local microbiologic flora and resistance patterns.

Table 25.1 Optimal timing for antibiotic initiation for patients with different severities of infection

Within the first hour of diagnosis	**Suspected infection** in patient with hemodynamic instability, neuromeningeal symptoms, neutropenia or post-splenectomy
Within the first 6 h of diagnosis	**Severe infection** in a stable patient without neuromeningeal symptoms, neutropenia or post-splenectomy
Within the first 24 h of diagnosis	**Non severe infection** in a stable patient without without neuromeningeal symptoms, neutropenia or post-splenectomy

Modified with permission from Wolters Kluwer Health; Antibiotic therapy in patients with septic shock. Eur J Anaesthesiol. 2011;28(5):318–24

Combination Antibiotic Therapy

Combination therapy is sometimes required as initial empirical therapy in order to cover multiple possible pathogens commonly associated with the specific infection. Frequently combination therapy with different modes of action are used (eg., beta lactam/aminoglycoside, beta lactam/fluroquinolone, beta lactam/macrolide-clindamycin combinations). If one of the antibiotics is already sensitive, the incremental role of the second antibiotic that is also sensitive is uncertain. Combination therapy has been shown to be beneficial in animal models, and in certain diseases such as endocarditis and cryptococcal meningitis. However meta-analyses have failed to demonstrate evidence of benefit for combination therapy.

De-escalation of Antibiotic Therapy

As a consequence of high mortality associated with inappropriate initial empiric therapy, these antibiotics have to cover a broad spectrum of pathogens. Continued use of such broad spectrum antibiotics predisposes to recurrence of infections with multi resistant pathogens in the same patient, increases the risk of *clostridium difficile* infection and contributes to general increase in resistance problems in the unit as a whole. So the empiric broad spectrum therapy should be aggressively de-escalated within 48–72 h if a plausible pathogen is identified. This can be achieved by changing the antimicrobial to one that has narrower spectrum of cover or by converting combination therapy to monotherapy. In patients who are culture negative, de-escalation is still possible after the patient stabilises clinically (i.e., after shock resolution). Such an approach will maximise the chances of achieving appropriate empirical antimicrobial therapy while minimizing selection pressure for resistant organisms.

Duration of Antibiotic Therapy

Duration of antimicrobial therapy has traditionally been based on expert opinion. With the increased awareness of the harmful effects of prolonged courses of antimicrobials, current trend is to give shorter courses of treatment. For example, for ventilator associated pneumonia, an 8 day course is as effective as 15 day course. Patients on shorter course of antibiotics develop fewer recurrent infections with multi-resistant pathogens. However, in some conditions such as endocarditis, osteomyelitis and intra-abdominal abscesses, longer regimes of 4–6 weeks are needed. In many such situations, the treatment duration has to be individualised after carefully considering the clinical and radiological response.

Dose Optimization in Critically Ill Patients

It is important to understand the many factors that affect the amount of antibiotics arriving at the target tissue in the acutely ill patients (see Chap. 6). Adjustment of doses and choices of antibiotics with these factors taken into consideration will more often result in a better outcome.

Antibiotic Classification in Sepsis

Antibiotics can be classified by their affinity for water to hydrophilic and lipophilic antibiotics (Table 25.2). Knowing this will allow the provider to understand how far the antibiotic will reach in the various types of tissues. They can also be classified by their pharmacodynamics into concentration-dependent antibiotics, time-dependent antibiotics and concentration-dependent antibiotics with time dependence (Fig. 25.1). These will give rise to apparent volume of distributions (V_d) which are different for various antibiotics.

Hydrophilic antibiotics are well distributed in both the intravascular and inter-stitial space (the extracellular space) and are not available in the intracellular space in meaningful concentrations. So their volume of distribution (V_d) is equivalent to the extracellular water.

Lipophilic antibiotics are available both in the intracellular space and in adipose tissues since they can cross lipid membranes. So their V_d is dependent on the amount of adipose tissue which is in turn dependent on the total body weight of the individual.

Table 25.2 Types of antibiotics based on affinity for water

Types of antibiotics	Pharmacokinetics	
	In healthy individuals	In patients with severe sepsis
Hydrophilic antibiotics e.g. beta lactams, aminoglycosides, glycopeptides, polymyxin	Limited intracellular penetration Low V_d Predominantly renal elimination	Increased V_d resulting in decreased plasma concentration Clearance increased if augmented renal clearance or decreased if renal impairment
Lipophilic antibiotics e.g. fluoroquinolones, macrolides, linezolid, tigecycline	High intracellular penetration Large V_d Elimination by hepatic metabolism	Minimal change in V_d Clearance dependent on hepatic function

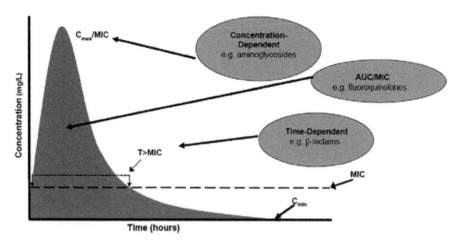

Fig. 25.1 Types of antibiotics based on pharmacokinetics of the drugs (With permission from Wolters Kluwer Health Critical Care Medicine 2008; 36(8):2433–40)

For concentration-dependent antibiotics the optimal activity correlates with ratio of peak serum concentration (Cmax) with minimum inhibitory concentration (MIC) of the bacteria (Cmax/MIC) (Fig. 25.1). For time-dependent antibiotics, the optimal activity correlates with the proportion of time the dosing interval the unbound drug (fT) is maintained above the MIC of the bacteria ($fT > MIC$). For concentration-dependent antibiotics with time dependence, it is the ratio between unbound fraction area under the curve of the drug ($fAUC_{0-24}$) and the MIC of the bacteria ($fAUC_{0-24}/MIC$).

Pharmacokinetic and Pharmacodynamic Considerations

Dosing

The V_d in critically ill patients with sepsis is increased due to the capillary leak syndrome which results in the shifting of fluids to the interstitial space. In critically ill patients, the V_d is also increased secondary to mechanical ventilation, fluid resuscitation, post-surgical drains, hypoalbuminemia (which causes plasma leakage) and extracorporeal circuits such as cardiopulmonary bypass. The increase in V_d has a dilutional effect on hydrophilic drugs resulting in lowering of their plasma concentration. This dilution effect has little effect on lipophilic antibiotics. However obesity increases the V_d of lipophilic drugs.

Optimal loading dose (LD) depends on V_d and target plasma concentration (C_t) and is calculated using the formula $LD = V_d \times C_t$. Therefore in severe sepsis where the V_d is increased, a higher loading dose of hydrophilic antibiotics is required to achieve the required target concentration ($C_t = LD/VD$). This is especially important for concentration dependent antibiotics such as aminoglycosides.

Clearance

While the V_d has a bearing on the initial dosing requirements, the clearance of the drug affects the daily dosing requirements to maintain steady state. A sizeable proportion of critically ill patients have higher than normal cardiac indices, resulting in increased renal perfusion. This leads to augmented renal clearance and increased elimination of hydrophilic drugs. The resulting shortened half-life necessitates more frequent dosing or administration as continuous infusions in order to increase the $fT > MIC$ ratio. Patients at risk in this category include the young, those following trauma, post operative situations and those with low illness severity scores. On the other hand, organ dysfunction can reduce clearance of the drug resulting in accumulation, so smaller doses or less frequent dosing will be necessary.

Hypoalbuminemia

Up to 40–50 % of critically ill patients can have low serum albumin. Hypoalbuminemia affects the pharmacokinetics of highly protein bound antibiotics such as ceftriaxone (85–95 % protein bound), cloxacillin (95 %), ertepenam (85–95 %) and teicoplanin (90–95 %). Low albumin levels lead to decreased protein binding and increased free (unbound) fraction. Since the unbound drug can freely diffuse into the extracellular space, there is a resultant increase in Vd. Also the unbound drug is cleared more easily by the kidney. Hypoalbuminemia therefore increases the Vd and clearance of highly protein bound antibiotics, resulting in sub optimal treatment when normal doses are used and an attempt should be made to increase the doses of these drugs or to increase the dose frequency.

Renal Dysfunction

Hydrophilic antibiotics are cleared mainly by the kidney and therefore tend to accumulate in the presence of renal dysfunction. Dose adjustments could be in the form of dose reduction or extended dosing intervals. For time-dependent antibiotics such as beta lactams, the more appropriate method would be to reduce the dose rather than the frequency of administration so that the $fT > MIC$ is preserved. For concentration dependent antibiotics such as aminoglycosides, the preferred method is to prolong the interval between the doses rather than to decrease the dose so that the Cmax/MIC ratio is preserved. Therapeutic drug monitoring (TDM) is frequently used for aminoglycosides and glycopeptides to fine tune the dosing regimes.

Table 25.3 summarises the general guidelines for antibiotic dosing in the critically ill septic patients.

Table 25.3 General guidelines for antibiotic dosing in critically ill septic patients

Specific antibiotics	Type of antibiotic	Protein binding	Dosing in critically ill with sepsis	Dosing in renal dysfunction	Dosing in hepatic dysfunction
Beta lactams e.g. penicillins, cephalosporins, carbapenams	Hydrophilic Time-dependent	Low except ceftriaxone, cloxacillin, ertepenam	High loading dose on day 1 (increased V_d) Frequent dosing or extended infusions (increased clearance)	Reduce dose rather than the frequency	Normal dosing
Glycopeptides e.g. vancomycin, teicoplanin	Hydrophilic Concentration dependent with time dependence	Vancomycin 30–55 % Teicoplanin 90 %	High loading dose on day 1 (increased V_d)	Titrate dose according to TDM results	Normal dosing
Aminoglycosides e.g. gentamicin, amikacin	Hydrophilic Concentration-dependent	Low	High loading dose on day 1 (increased V_d)	Prolong interval between doses. Titrate dose according to TDM	Normal dosing
Fluoroquinolones e.g. ciprofloxacin, levofloxacin, moxifloxacin	Lipophilic Concentration dependent with time dependence	20–52 %	Normal dosing	Dose adjustments required for ciprofloxacin and levofloxacin	Normal dosing

Key Concepts

- Early initiation of appropriate antimicrobial therapy is of paramount importance to improve survival.
- The initial regime has to be of broader spectrum especially in patients with risk factors for infection with multiresistant pathogens.
- Pharmacokinetic profiles of antibiotics are modified in critically ill patients.
- In critically ill patients with no organ impairment, a loading dose is advisable when beta lactams, aminoglycosides or glycopeptides are being used.
- Once culture results are available aggressive de-escalation has to be done to prevent the adverse effects of prolonged administration of broad spectrum agents.
- Duration of treatment has to be individualised, with increased emphasis on shorter treatment regimes.

Summary

Combating sepsis involves not only choosing the correct antibiotics but also the adjustment of doses or frequency of antibiotic administration so that the blood or tissue level of the antibiotic is appropriate. Understanding the changes in pharmacokinetics and pharmacodynamics of antibiotics in the setting of the critically ill patient is a useful first step in the proper management of this challenging problem.

Further Reading

1. Beardsley JR, Williamson JC, Johnson JW, Ohl CA, Karchmer TB, Bowton DL. Using local microbiologic data to develop institution-specific guidelines for the treatment of hospital-acquired pneumonia. Chest. 2006;130(3):787–93.
2. Chastre J, Wolff M, Fagon JY, Chevret S, Thomas F, Wermert D, et al. Comparison of 8 vs 15 days of antibiotic therapy for ventilator-associated pneumonia in adults: a randomized trial. JAMA. 2003;290(19):2588–98.
3. Claus BO, Hoste EA, Colpaert K, Robays H, Decruyenaere J, De Waele JJ. Augmented renal clearance is a common finding with worse clinical outcome in critically ill patients receiving antimicrobial therapy. J Crit Care. 2013;28(5):695–700.
4. Garnacho-Montero J, Garcia-Garmendia JL, Barrero-Almodovar A, Jimenez-Jimenez FJ, Perez-Paredes C, Ortiz-Leyba C. Impact of adequate empirical antibiotic therapy on the outcome of patients admitted to the intensive care unit with sepsis. Crit Care Med. 2003;31(12):2742–51.
5. Kumar A, Ellis P, Arabi Y, Roberts D, Light B, Parrillo JE, et al. Initiation of inappropriate antimicrobial therapy results in a fivefold reduction of survival in human septic shock. Chest. 2009;136(5):1237–48.
6. Moore RD, Lietman PS, Smith CR. Clinical response to aminoglycoside therapy: importance of the ratio of peak concentration to minimal inhibitory concentration. J Infect Dis. 1987;155(1):93–9.
7. Paul M, Benuri-Silbiger I, Soares-Weiser K, Leibovici L. Beta lactam monotherapy versus beta lactam-aminoglycoside combination therapy for sepsis in immunocompetent patients: systematic review and meta-analysis of randomised trials. BMJ (Clinical research ed). 2004;328(7441):668.
8. Roberts JA, Kruger P, Paterson DL, Lipman J. Antibiotic resistance–what's dosing got to do with it? Crit Care Med. 2008;36(8):2433–40.
9. Safdar N, Handelsman J, Maki DG. Does combination antimicrobial therapy reduce mortality in Gram-negative bacteraemia? A meta-analysis. Lancet Infect Dis. 2004;4(8):519–27.
10. Singh N, Rogers P, Atwood CW, Wagener MM, Yu VL. Short-course empiric antibiotic therapy for patients with pulmonary infiltrates in the intensive care unit. A proposed solution for indiscriminate antibiotic prescription. Am J Respir Crit Care Med. 2000;162 (2 Pt 1):505–11.
11. Textoris J, Wiramus S, Martin C, Leone M. Antibiotic therapy in patients with septic shock. Eur J Anaesthesiol. 2011;28(5):318–24.
12. Ulldemolins M, Roberts J, Rello J, Paterson D, Lipman J. The effects of hypoalbuminaemia on optimizing antibacterial dosing in critically ill patients. Clin Pharmacokinet. 2011;50 (2):99–110.

Chapter 26
The Allergic Patient

Sook Hui Chaw and Kit Yin Chow

Abstract Allergy refers to an exaggerated response by our immune system to exposure to certain allergens which are normally harmless. Allergy manifestations can range from mild symptoms to a life-threatening event related to anaphylaxis, also known as a type 1 hypersensitivity reaction. The cornerstone of management is strict avoidance of known allergens, prompt diagnosis and pharmacological treatment to counteract the actions of allergy mediators. Adrenaline is the recommended first-line treatment in anaphylaxis. Second-line pharmacological agents include H1-antihistamines, corticosteroids, beta 2-agonists and vasopressors. In severe anaphylaxis, support of the cardiovascular system with oxygen and fluids as well as adrenaline may be necessary. Failure to relieve bronchospasm by pharmacological means may require mechanical ventilation to maintain effective oxygenation.

Keywords Allergy • Anaphylaxis • Hypersensitivity reaction • Treatment of anaphylaxis • Adrenaline

Introduction

Allergy refers to an exaggerated response by our immune system to exposure to certain foreign substances (allergens) which are normally harmless. Allergy is one of the four forms of hypersensitivity and is called type 1 or immediate hypersensitivity. Allergic reactions can result from airborne particles such as dust or pollen, foods, insect stings and medications.

Allergic reactions range from mild to life threatening. Mild allergic symptoms include runny nose, sneezing, watery eyes or nasal congestion. However, exposure to some allergens can lead to severe anaphylaxis that can affect multiple organ systems and may cause death.

S.H. Chaw, M.D., M.Anaes (✉) • K.Y. Chow, M.B.B.S., M.Anaes
Department of Anesthesiology, Faculty of Medicine, University of Malaya,
50603 Kuala Lumpur, Malaysia
e-mail: sh_chaw@um.edu.my

© Springer International Publishing Switzerland 2015
Y.K. Chan et al. (eds.), *Pharmacological Basis of Acute Care*,
DOI 10.1007/978-3-319-10386-0_26

Pathophysiology

Anaphylaxis is also known as a type I hypersensitivity reaction. There is an initial exposure to a foreign protein or antigen, which stimulates the production of IgE antibodies. The antibody concentrations decrease after the initial exposure but IgE binds to the mast cells and basophils. With further exposure, the antigen will bind to the IgE antibodies and result in the release of mediators such as histamine, tryptase and prostaglandins.

Anaphylactoid reactions are not IgE mediated and do not require previous exposure to the antigen. Mechanisms involve the release of vasoactive substances e.g. histamine, direct histamine release from mast cells, or complement activation.

Both anaphylactic and anaphylactoid reactions are dose independent and are indistinguishable from each other clinically. Clinical manifestations include increased mucous secretion, bronchospasm, increased vascular permeability, airway edema, and hypotension.

Non-immunological histamine release is the result of direct action of a drug on mast cells. The degree of clinical response is dependent on the drug dose and delivery rate. It is usually benign with only skin involvement.

Drugs Used in the Management of Allergic Reactions

The mainstay of management in allergy is strict avoidance of allergens and pharmacological treatment to block the actions of allergic mediators, or to prevent activation and degranulation of mast cells. These medications alleviate symptoms and are important in recovery from acute anaphylaxis.

Adrenaline

Adrenaline is very important in the treatment of anaphylaxis due to its pharmacologic effects through α- and β- receptors. The mechanism of action through the receptors is different (Table 26.1).

Table 26.1 Mechanism of action at the alpha 1 and beta 2 -receptors

Mechanism of action	
Alpha 1-receptor	Beta 2-receptor
1. Increase vasoconstriction --> Increase blood pressure	1. Increase bronchodilation -> relieves wheezing
2. Decrease laryngeal edema -> relieves upper airway obstruction	2. Decrease mediator release -> decrease hives

Other pharmacologic effects through $beta_1$- receptors: Increase heart rate and force of cardiac contractions.

For all cases of anaphylaxis, the World Allergy Organisation and the National Institutes of Health-National Institute of Allergy and Infectious Diseases (NIH-NIAID) food allergy guidelines recommend adrenaline as the first-line of treatment. Early diagnosis of anaphylaxis and administration of adrenaline in adequate dosage is important to optimize outcome. The World Allergy Organization has recognized that adrenaline administration was often delayed or inadequate in anaphylactic fatalities.

Adrenaline Dosage and Route of Administration

Intramuscular (IM) adrenaline should be considered in the initial management when anaphylaxis is diagnosed, or strongly suspected, or when intravenous (IV) access is not established. It should be injected at mid-anterolateral thigh in a dose of 0.01 mg/kg of 1:1,000 (1 mg/ml) solution, to a maximum of 0.5 mg in adults (0.3 mg in children). The dose can be repeated every 5–15 min depending on the response and the severity of anaphylaxis.

When IV access is established and in the presence of close hemodynamic monitoring, adrenaline 1:10,000 (0.1 mg/ml) can be administered intravenously in small doses, titrating to effect (50 mcg (0.5 ml) boluses in adults, 1 mcg/kg in children).

Adverse Effects of Adrenaline

Serious adverse effects such as hypertensive crises, ventricular arrhythmias and pulmonary edema can occur after an overdose of adrenaline. Groups of patients who are at higher risk in the event of overdose include patients at the extremes of age and patients with ischemic heart disease, hypertension or hyperthyroidism. Cocaine and amphetamines sensitize the myocardium to the effects of adrenaline. In contrast, beta blockers and angiotensin converting enzyme inhibitors can decrease the effectiveness of endogenous and exogenous catecholamines.

H1-Antihistamines

H1-antihistamines are a second-line medication for anaphylaxis and should not be a substitute for adrenaline. In Cochrane systematic review, the administration of H1-antihistamine in anaphylaxis was not supported by evidence from randomized controlled trials.

Examples of H1-antihistamines are IV Chlorpheniramine (10–20 mg in adults; 0.2 mg/kg in children) or IV diphenhydramine (10–50 mg in adults; 5 mg/kg/24 h in children). H1-antihistamines exert their pharmacologic effects at H1-receptors with inverse agonist effect and stabilize receptors in an inactive conformation. Compared to adrenaline, H1-antihistamines have slow onset of action.

H1-antihistamines are useful for symptomatic treatment of flushing, urticaria, pruritus, sneezing and rhinorrhea but are not life-saving as they have little effect on bronchospasm and hypotension.

First-generation H1-antihistamines can cause somnolence and impaired cognitive function with the usual dosage. In the event of overdosage, they can cause extreme drowsiness, coma, respiratory depression and seizures in infants and children.

Corticosteroids

Hydrocortisone is the most commonly used corticosteroid in the event of an anaphylactic or allergic reaction. This drug inhibits the production of inflammatory cytokines. As a result, this will decrease inflammation, cellular edema and cellular recruitment at affected sites.

However, it should be noted that the onset of the anti-inflammatory effect produced by corticosteroids is delayed and the duration of action is 8–12 h. The beneficial effect in anaphylaxis or allergic reactions is more likely due to suppression of the anti-inflammatory response rather than to the inhibition of the production of immunoglobulins. Therefore, the mainstay of treatment in the event of life-threatening anaphylaxis is still adrenaline.

Hydrocortisone is administered intravenously 200 mg immediately, and followed by 4–6 hourly doses.

Beta-2 Agonists

Selective beta 2 adrenergic agonists are sometimes given as second-line medication in anaphylaxis for treating bronchospasm through its effect on beta-2 receptors in bronchial smooth muscle. This class of drugs has minimal alpha-1 adrenergic agonist effects and therefore should not be a substitute for adrenaline. Salbutamol is a beta-2 agonist that is administered by inhalation – 0.5 ml of 0.5 % solution (2.5 mg) nebulized over 20 min.

Potential adverse effects of beta-2 agonists are tremor, tachycardia, headache, dizziness, vasodilation and hypokalemia.

Vasopressors

These can be divided into catecholamines and non-catecholamines. Commonly used catecholamines include noradrenaline and dopamine whereas ephedrine and phenylephrine are the two most commonly used non-catecholamines.

This group of drugs is the second-line treatment in anaphylaxis or allergic reactions. They are usually given when the patient remains hemodynamically unstable despite adrenaline and adequate fluid resuscitation.

Direct-acting sympathomimetics include cathecholamines and phenylephrine (non-catecholamine). These drugs act by activating adrenergic receptors directly.

Ephedrine is an indirect-acting non-catecholamine. It evokes the release of the endogenous neurotransmitter norepinephrine from the postganglionic sympathetic nerve endings, which will in turn activate the adrenergic receptors.

Key Concepts

- Allergy refers to an exaggerated response by our immune system to exposure to certain foreign substances which are normally harmless.
- The mainstay of management in allergy and anaphylaxis is strict avoidance of allergens, early clinical recognition of allergic reaction and prompt treatment.
- Pharmacological treatment alleviates symptoms of allergic reaction by blocking the actions of allergic mediators, or preventing activation and degranulation of mast cells.
- Adrenaline is the first-line of treatment in anaphylaxis. Prompt administration of adrenaline in adequate dosage is important to optimize outcome.

Summary

Allergy refers to an exaggerated response by our immune system to a foreign antigen. Allergic reactions range from mild to a life threatening form, called anaphylaxis. Patients with allergies should strictly avoid known allergens. Treatment of an allergic reaction is mainly supportive and pharmacological agents are used to block the actions of the chemical mediators. The drug of choice for an anaphylactic reaction is adrenaline which stimulates alpha-1 and beta- 2 receptors. Care should be taken to avoid overdose and serious adverse effects such as hypertensive crisis, arrhythmias and pulmonary edema. Antihistamines such as H1-antagonists can be used as adjunct treatment and should not be used alone in an anaphylactic reaction. Corticosteroids such as hydrocortisone are efficacious in reducing the inflammatory response and cellular edema. Beta-2 agonists can also be administered by inhalation to treat bronchospasm associated with an allergic reaction. Vasopressors may be considered in an anaphylactic reaction when there is suboptimal response to adrenaline and fluid resuscitation.

Further Reading

1. Boyce JA, Assa'ad A, Burks AW, Jones SM, Sampson HA, Wood RA, et al. Guidelines for the diagnosis and management of food allergy in the United States: report of the NIAID-sponsored expert panel. J Allergy Clin Immunol. 2010;126(6 Suppl):S1–58.
2. Lieberman P, Nicklas RA, Oppenheimer J, Kemp SF, Lang DM, Bernstein DI, et al. The diagnosis and management of anaphylaxis practice parameter: 2010 update. J Allergy Clin Immunol. 2010;126(3):477–80, e1–42.
3. Schwartz LB. Systemic anaphylaxis, food allergy, and insect sting allergy. In: Goldman L, Schafer AI, editors. Cecil Medicine. 24th ed. Philadelphia: Saunders Elsevier; 2011.
4. Simons FE, Ardusso LR, Bilo MB, El-Gamal YM, Ledford DK, Ring J, et al. World Allergy Organization anaphylaxis guidelines: summary. J Allergy Clin Immunol. 2011;127(3):587–93, e1–22.

Chapter 27
The Poisoned Patient

Choo Hock Tan and Ahmad Khaldun Ismail

Abstract Acute exposure to toxic or poisonous substances can be intentional or accidental. The onset of the toxic effect depends on the degree of absorption of the toxicant. The management of acute exposure to toxicants is generally supportive. Knowledge of the pharmacokinetics and pharmacodynamics of the offending agent is important. The use of pharmacological agents in treating poisoning is to manipulate the pharmacokinetic or pharmacodynamic profiles of the toxicant. These pharmacological agents will minimize the absorption and/or enhance the elimination of the toxicant from the body. Specific antidotes are used to alter the effects of certain known toxicants through neutralization and pharmacological antagonism. Having the ability to identify the effect of specific classes of substances on the body (toxidromes) will facilitate the selection of an appropriate management strategy to optimize the outcome.

Keywords Antagonism • Antidote • Toxicant • Toxin • Poison • Medical toxicology • Neutralization • Rebound phenomenon

Introduction

Medical toxicology is one of the four subspecialties of toxicology. It deals with human exposure to toxic substances. The origin of these toxic substances can be from animal, plant or synthetic materials. Human exposure to these toxoids and toxins can either be intentional (deliberate), unintentional (accidental) or fabricated (induced). Many toxins and toxoids have long been utilized as pesticides, herbicides, fertilizers, household cleaners, cosmetics, pharmaceutical and other useful

C.H. Tan, M.B.B.S., Ph.D. (✉)
Department of Pharmacology, Faculty of Medicine, University of Malaya,
50603 Kuala Lumpur, Malaysia
e-mail: tanch@um.edu.my

A.K. Ismail, MBBCh, BAO, BMedSc, Dr.Em.Med
Department of Emergency Medicine, Faculty of Medicine, UKM Medical Centre, Jalan
Yaacob Latif, Bandar Tun Razak, 56000 Cheras, Kuala Lumpur, Malaysia

© Springer International Publishing Switzerland 2015
Y.K. Chan et al. (eds.), *Pharmacological Basis of Acute Care*,
DOI 10.1007/978-3-319-10386-0_27

products of daily use. Unfortunately, some of these substances have been used for intentional poisoning (of individuals, targeted groups of people or the general public), biological warfare and antipersonnel terrorism.

General Principles

The effects of poisoning or envenoming begin with the exposure to large enough amounts of xenobiotics (foreign, natural or man-made substances) to cause harm or damage to the body. These toxic substances enter the body through various routes such as oral (most common), intradermal, intravascular or inhalational. The rate of absorption predicts the onset of action, while the extent of absorption reflects the bioavailability and its pharmacological effect. Absorption rate, bioavailability and elimination of these toxicants are influenced by physical factors (molecular weight, blood flow, surface area, contact time) and chemical factors (water/lipid solubility, polarity, pH of the medium). The general approach to toxic exposures is to remove the patient from the substance and the substance from the patient as soon as and as completely as possible.

Approaches to Acute Poisoning

This involves various methods and pharmacological manipulation to alter the pharmacokinetics and pharmacodynamics of toxicants. Treatment should take place in tandem with but not be delayed by diagnostic, screening or laboratory investigations.

1. **Preventing further gastrointestinal (GI) absorption (Gastric decontamination)**

   (a) Emesis: Syrup ipecac (active compounds: emetine and cephaeline) act locally by irritating the gastric mucosa and centrally by stimulating the medullary chemoreceptor trigger zone to induce vomiting to reduce the poison load. Ipecac is only considered immediately post toxic ingestion due to the significant risk of aspiration especially in those with decreased level of consciousness.

   (b) Activated charcoal: This is the most frequently used agent for GI decontamination. The high degree of microporosity (>500 m^2 per g) of activated charcoal makes it an excellent adsorbing agent, binding to a wide selection of drugs and their metabolites. When administered early following toxic ingestion, it significantly reduces gastrointestinal absorption, enterohepatic circulation and enteroenteric circulation.

   (c) Cathartic: Cathartics enhance bowel transit. The indication is similar to activated charcoal and is often used with activated charcoal. Magnesium citrate, magnesium sulfate and sorbitol can decrease intestinal transit time and absorption of the ingested xenobiotics. Severe diarrhea is a major adverse effect and requires careful monitoring of the hydration status. Magnesium salt-containing cathartics should be avoided in patients with renal failure, due to risk of magnesium toxicity.

(d) Gastrointestinal irrigation: Orogastric lavage with small amounts of warm fluid through an appropriately placed orogastric tube is only considered in alert patients presenting within 60 min of life threatening toxic ingestions. This procedure is limited to the irrigation of preparations small enough to pass through the orogastric tube, and substances that cause delay in gastric emptying, gastric outlet obstruction or concretion. Whole bowel irrigation using non-absorbable polyethylene glycol has a role in the toxic ingestion of petroleum products, iron, lithium, sustained-release-enteric-coated formulations or in patients who are "body packers" i.e. transporting illicit drugs in their bowel.

Clinical Correlation – Caution with Gastric Lavage and Removal from the Lower Gastrointestinal Tract

Inducing emesis or orogastric lavage involves manipulation of the upper gastrointestinal tract. This should be avoided in patients with decreased level of consciousness, seizure, absent gag reflex, and caustic ingestion. While activated charcoal works well for most drugs (aspirin, carbamazepine, digitalis, barbiturates, phenytoin etc.), it is not useful for substances like petroleum products, corrosives, mineral acids, ethanol/glycols, cyanide, boric acid, lithium, iron, and certainly must not be given with oral antidotes as it can bind and inactivate these agents. Manipulation of the lower intestinal tract (laxative use, bowel irrigation) is generally contraindicated in gut atony-related conditions and comatose or convulsive patients due to risk of gut perforation and aspiration.

---

2. **Increasing elimination (by manipulating urine pH, or by extracorporeal techniques)**

(a) Alkalinization of urine – weak acids like salicylates and barbiturates, ionize in alkaline urine resulting in an increase in their renal excretion (Fig. 27.1). Sodium bicarbonate is administered as IV bolus or infusion to maintain the urinary pH between 7.5 and 8.5. Continuous infusion is adjusted to clinical response or until serum pH is maintained between 7.50 and 7.55. This procedure is contraindicated in hypokalemia, and in those with renal insufficiency (not able to tolerate volume or sodium load).

(b) Acidification of urine – Basic drugs like amphetamines, quinine, ephedrine and flecainide, ionize in acidic urine and are excreted more readily. Urine acidifiers include ammonium chloride and vitamin C.

(c) Non-pharmacological approaches – peritoneal dialysis or hemodialysis can increase the elimination of salicylates, lithium, barbiturates, methanol, ethylene glycol and ethanol.

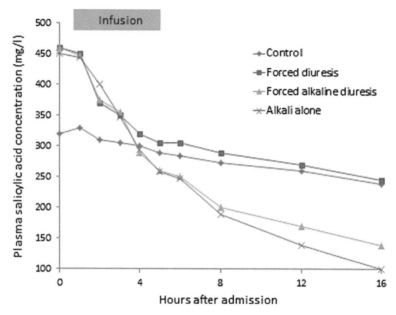

Fig. 27.1 Effects of different treatment regimens (fluid and alkali) on the mean plasma concentrations of salicylic acid in patients with aspirin overdose. Alkalinising the urine promotes ionisation of salicylic acid, reducing salicylate reabsorption and hastening its elimination from the body (Permission from BMJ Publishing Ltd; Br Med J (Clin Res Ed) 1982;285(6352):1383–6)

Clinical Correlation – Caution with Diuresis

In some settings, diuretics are administered to increase the rate of toxicant excretion. Where forced alkaline diuresis is indicated, it is important to first correct plasma volume depletion, electrolyte and metabolic abnormalities, as well as to ensure the renal function is normal before commencing the procedure. Sodium bicarbonate alone is often effective without forced diuresis (Fig. 27.1). Theoretically, urine acidification can promote the excretion of basic drugs, but this procedure is not recommended because it can cause metabolic acidosis.

---

3. **Use of antidote (interfering with the pharmacodynamics or toxic effects of poison)**

   (a) Neutralization of the circulating toxicant/toxin – this can be achieved with the use of antibody e.g. DigiFab (for digoxin) and antivenom (for venom), or chelating agents such as desferrioxamine (for iron). The rationale is that the therapeutic agent binds (chemically or immunologically) to the toxic agent, rendering it inactive while enhancing its elimination (e.g. immunocomplexes will be phagocytozed).

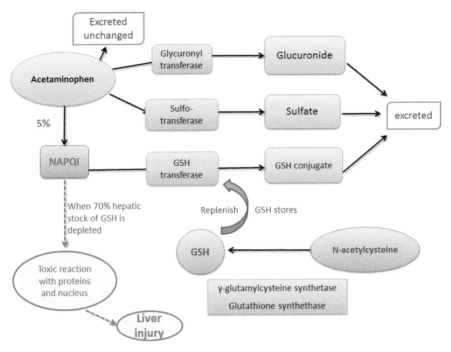

Fig. 27.2 In a healthy adult, almost 95 % of acetaminophen is conjugated with glucuronide and sulfate (Phase II reaction), followed by excretion in the urine. The small amount of hepatoxic metabolite, N-acetyl-p-benzoquinone imine (NAPQI), is usually conjugated to glutathione (GSH) for excretion. In acetaminophen overdose, the glutathione stock is depleted, preventing conjugation and resulting in the accumulation of NAPQI that leads to cell necrosis and liver failure. The antidote, N-acetylcysteine, when administered, serves to replenish the GSH for the detoxification of NAPQI

(b) Antagonizing the toxic effects – typically this can be achieved with the use of an antagonist, such as atropine (for cholinergic poisoning), and cholinesterase inhibitor (for non-depolarizing neuromuscular blocker overdose). In organophosphate poisoning that inhibits cholinesterase, pralidoxime is an antidote that reactivates the phosphorylated cholinesterase. Other examples include flumazenil and naloxone that competitively antagonize the actions of benzodiazepines and opioids, respectively (see Chap. 7).

(c) Reducing the generation of toxic metabolite: This is useful for drugs which have toxic metabolites for example, acetaminophen (paracetamol). N-acetylcysteine is an antidote that restores hepatic glutathione, which detoxifies the toxic metabolite, *N*-acetyl-p-benzoquinone imine (NAPQI) (Fig. 27.2).

Clinical Correlation – Beware of Resurgence of the Toxic Effects

The use of antidote to bind circulating antigens (e.g. digoxin, venom toxins) requires careful assessment for rebound phenomenon, especially for drugs and toxins that distribute widely into the tissues (recall: digoxin V_d = 500 L). Immunocomplexation enhances the clearance of the drug / toxin from the vascular compartment, resulting in a sudden shift of the intercompartmental equilibrium and a subsequent resurgence of antigens from the peripheral compartments (tissue-deposited drugs) into the vascular compartment. In the case where a substantial amount of the antidote has been cleared, this can be accompanied with the reappearance of clinical signs, needing prompt management and additional doses of the antidote. This is seen in some snake envenomation where there is a pharmacokinetic-pharmacodynamic mismatch between the venom and antivenom, indicating that the clinical effectiveness of the antivenom needs to be further optimized (Fig. 27.3).

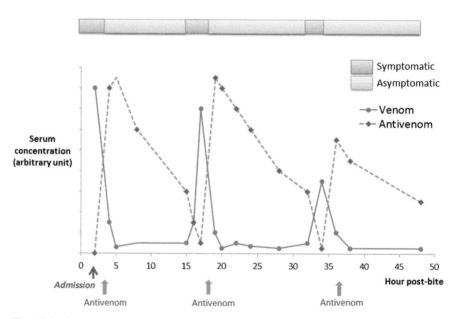

Fig. 27.3 Conceptual representation of venom (*unbroken line*) and antivenom (*broken line*) levels in a patient following snake envenomation. Note the toxic effect subsides (patient becomes asymptomatic, indicated by *green bar*) as the venom level is depleted by antivenom. Rebound phenomenon occurs when the venom level resurges as the antivenom level decreases. This is often accompanied with recurrence of clinical syndrome (e.g. recurrent coagulopathy, recurrent paralysis), necessitating further doses of antivenom to ensure sustained and complete neutralization

Key Concepts

- Drug manipulation can alter the pharmacokinetics (reducing absorption or enhancing elimination) and pharmacodynamics (preventing or ameliorating toxic effect) of toxic substances.
- Choosing the appropriate method of treating poisoning requires good understanding of how different toxic substances are introduced and act on the body. The dose and duration of exposure prior to receiving medical care will determine the appropriate treatment pathway.

Summary

The primary goal in the management of poisoning is to prevent or reduce further injury. In general, supportive care plays a major role in all types of poisoning. Pharmacological approaches, however, are useful to reduce the absorption and to enhance the elimination of the toxic substance. The use of appropriate antidote may prove to be a useful mode of therapy to neutralize the toxic effects of some compounds. Timely administration of appropriate treatment modalities for the acutely poisoned ensures a favorable outcome.

Further Reading

1. Brunton L, Blumenthal D, Buxton I, Parker K, editors. Goodman & Gilman manual of pharmacology and therapeutics. 11th ed. New York: McGraw Hill; 2008.
2. Harvey RA, Clark MA, Finkel R, Rey JA, Whalen K, editors. Lippincott illustrated reviews pharmacology. 5th ed. Philadelphia: Wolters Kluwer; 2012.
3. Hoffman RS, Howland MA, Lewin NA, Nelson LS, Goldfrank LR. Goldfrank's toxicologic emergencies. 10th ed. New York: McGraw-Hill Professional; 2014.
4. Marx J, Hockberger R, Walls R. Rosen's emergency medicine – concepts and clinical practice: expert consult premium edition. 8th ed. Philadelphia: Saunders; 2013.
5. Prescott LF, Balali-Mood M, Critchley JA, Johnstone AF, Proudfoot AT. Diuresis or urinary alkalinisation for salicylate poisoning? Br Med J (Clin Res Ed). 1982;285(6352):1383–6.
6. Rang HP, Dale MM, Ritter JM, Flower RJ, Henderson G. Rang and Dale's pharmacology. 7th ed. Edinburgh: Churchill Livingstone; 2012.
7. Tintinalli JE, Stapczynski JS, Ma OJ, Cline DM, Cydulka RK, Meckler GD. Tintinalli's emergency medicine: a comprehensive study guide. 7th ed. New York: McGraw-Hill; 2010.
8. Tripathi KD. Essentials of medical pharmacology. 7th ed. New Delhi: Jaypee Brothers; 2013.
9. White J, Meier J. Handbook of clinical toxicology of animal venoms and poisons. 1st ed. Boca Raton: CRC Press; 1995.

Part IV
Pharmacology of Special Drugs

Chapter 28
Oxygen as a Drug

Yoo Kuen Chan

Abstract Oxygen is a drug naturally available in the atmosphere. In the acutely ill patient, we administer it either from the environment at 21 % or from stored oxygen sources in higher concentrations. It is administered via nasal prongs, masks or other oxygen delivery and/or ventilator devices. Oxygen is delivered to the tissues by the cardiopulmonary system and is mainly carried by hemoglobin. Under hyperbaric conditions, more oxygen is carried in solution in the blood. Oxygen is necessary for the optimal production of ATP within the cells, which is the source of energy for all cellular function, without which cell death occurs. It is administered whenever tissue oxygen supply may be compromised, such as in low inspired oxygen conditions, poor tissue perfusion states, and inadequate or impaired hemoglobin carrying capacity. Oxygen from hyperbaric sources is used to combat air embolism and also to assist in poorly healing wounds associated with cancer therapy. Hazards of oxygen therapy include retinopathy of prematurity, lung fibrosis and seizures. Oxygen administration areas are also fire risk zones.

Keywords Oxygen a drug • ATP production • Oxygen sources • Oxygen therapy • Hazards of oxygen

Introduction

A drug is a medicine or substance which has a physiological effect when ingested or otherwise introduced into the body. So oxygen can be defined as a drug, particularly in the ill patient, because oxygen is the most common substance we administer in acute care. It is not usually considered as a drug, because oxygen exists in the air that we breathe in (21 % or 150 mmHg pressure at sea level), and is not dispensed by the pharmacy, but rather by the engineering department of a hospital. In the normal state of health, this amount of oxygen in the atmosphere is more than

Y.K. Chan, M.B.B.S., FFARCS (Ireland) (✉)
Department of Anesthesiology, Faculty of Medicine, University of Malaya,
50603 Kuala Lumpur, Malaysia
e-mail: chanyk@um.edu.my

© Springer International Publishing Switzerland 2015
Y.K. Chan et al. (eds.), *Pharmacological Basis of Acute Care*,
DOI 10.1007/978-3-319-10386-0_28

sufficient to support us in a wide range of activities. However, in patients who are acutely ill, more oxygen than is available from the environment is needed to support the physiological processes in the body. In these situations, oxygen becomes a drug that we administer.

Pathophysiology

Humans being complex organisms need huge amounts of adenosine triphosphate (ATP) at the cellular level for all energy requiring activities. Without ATP, cell death occurs. ATP, which is the main source of energy at the cellular level, is produced from ingested energy sources according to the needs of the body – as demand for ATP increases, so does its production. ATP is not stored in significant amounts in the body. It is produced within the mitochondria by metabolism of energy substrates such as carbohydrates and fats. ATP production in the absence of oxygen (anaerobic metabolism), yields very much less ATP than that produced in aerobic conditions. The pathway or sequence of enzymatic reactions in which ATP is produced aerobically is called the Citric Acid Cycle or the Krebs cycle. In the presence of oxygen, 19 times more ATP is produced from one mol of glucose (Fig. 28.1) than without oxygen. This is the basic physiology behind the logic of ensuring that there is adequate oxygen available at the cellular level to allow for ATP production which will keep our cells, organs and ultimately our bodies, alive.

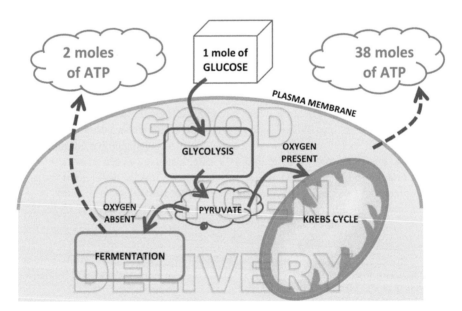

Fig. 28.1 ATP output from a mole of glucose is different in the presence or absence of oxygen at the cellular level

So, energy (ATP) requirements, and therefore oxygen requirements, increase as activity levels increase. A measure of the metabolic cost of activities can be described in terms of the metabolic equivalent or the Metabolic Equivalent of Task (MET). One MET is defined as the expenditure of 1 kcal/kg/h, which corresponds to oxygen requirement of approximately 3.5 ml/kg/min. One MET is the level of metabolism at rest or when sitting quietly – which is not the same as the basal metabolic rate. At 2 MET, twice the amount of energy is expended and oxygen requirement increases to 7 ml/kg/min. Light activities score <3 MET, moderate activities like brisk walking and bicycling MET 3–6 and vigorous activities (jogging) MET >6. Used clinically, the MET values are only estimates of level of activity, which also vary between individuals according to physiological factors such as age, weight and level of fitness.

Pharmacokinetics

Oxygen moves into the body by bulk flow, as part of the air we inhale enters the alveoli. Whilst breathing air, the oxygen tension in the alveoli is about 100 mmHg. This then diffuses across from the alveoli to the blood in the pulmonary capillaries. The pulmonary artery which brings in mixed venous blood (the collective venous return from all the organs in the body to the heart) to the alveoli, has an oxygen tension of only 40 mmHg. This allows a huge gradient for diffusion of oxygen across the alveolar-capillary membrane. The amount of oxygen transferred across from the lungs to the blood is dependent not only on the diffusion gradient but also on the match between the alveolar ventilation and the pulmonary blood flow. Oxygen is then moved by bulk flow in the blood to the tissues by the pumping action of the heart (called the cardiac output), mainly in combination with hemoglobin in red blood cells. Under normal atmospheric pressure the dissolved oxygen constitutes only 0.003 ml/mmHg PO_2 i.e. only about 0.3 ml of oxygen per 100 ml of blood, whereas the amount carried by the haemoglobin in a patient with a haemoglobin of 15 g% is about 19.5 ml per 100 ml of blood.

As oxygen delivery is mainly reliant on the level of hemoglobin, increasing this with the use of erythropoietin is a strategy to improve oxygen delivery in those with anemia. In a hyperbaric environment (usually 3 atmospheres), the amount of dissolved oxygen in the blood can be increased to 6 ml of oxygen per 100 ml of blood and this may be a method to increase tissue oxygen delivery in very unusual circumstances. At the tissue capillaries, oxygen diffuses from the capillary bed to the cells and the mitochondria. The oxygen tension in the mitochondria can be as low as 2–3 mmHg and yet support aerobic metabolism.

In those with severe life threatening respiratory dysfunction where the lungs can no longer sustain oxygen transfer, oxygen can be administered directly into the blood through the use of an Extracorporeal Membrane Oxygenation (ECMO) circuit or a cardiac bypass machine. This may also be used in patients with impossible to manage complete airway obstruction as a way of temporarily securing the delivery of oxygen to the tissues.

Sources of Oxygen

Oxygen can be obtained from the environment around us, from cylinders of compressed oxygen or from pipeline supplies drawn from liquid oxygen tanks, usually found in hospitals where usage is high.

Oxygen from the Environment

Oxygen from the air around us can be used to ventilate and oxygenate a patient with the use of self inflating ventilation devices like the Ambu Bag. In desperate situations when patients are unable to breathe adequately, these self inflating ventilation devices can be utilised to deliver air (21 % oxygen) to the lungs and can be life saving. Oxygen from the environment can be enriched from 21 % to nearly 100 % with the use of a device called an oxygen concentrator. Oxygen concentrators essentially pass air through zeolite which absorbs nitrogen, and concentrates the oxygen to a higher percentage. Oxygen concentrators are used for domiciliary purposes for patients with chronic conditions requiring long term oxygen therapy who can cope with care in the home environment.

Oxygen from Cylinders

Oxygen is stored in cylinders under pressure. The pressure of a full cylinder of oxygen is 134 atmospheres. Oxygen cylinders come in various sizes and smaller ones are used for transportation of patients requiring oxygen therapy. A regulator brings the pressure down to 4 atmospheres, after which the oxygen is passed through a flow meter or other equipment.

Liquid Oxygen

Liquid oxygen is stored in huge insulated cylinders at temperatures below zero degrees centigrade. Oxygen vaporises from the liquid state in these special cylinders and flows in the hospital pipelines to oxygen wall outlets. At these points, flow meters can be attached to enable delivery of oxygen directly to patients, or to equipment such as mechanical ventilators or anesthetic machines.

These liquid oxygen tanks are refilled from cylinders sent by trucks from the manufacturers.

Dosaging

Oxygen is administered as a percentage of oxygen in air or other carrier gas. It is described either as a percentage or fraction of inspired oxygen (FiO_2). So, the inspired oxygen concentration while breathing air is 21 % or FiO_2 is 0.21. If pure oxygen is administered, then the concentration is 100 % or FiO_2 is 1. Oxygen analysers are used to measure the concentration of oxygen delivered from mechanical ventilators and anaesthetic machines.

Methods of Administration

Oxygen can be administered via nasal prongs, oxygen masks, breathing systems, resuscitation devices and ventilators.

The flow meters allow a maximum oxygen delivery of 15 L/min. The rate of airflow into the upper respiratory tract during inspiration is around 30 L/min. Thus, administering 15 L/min of oxygen with an ordinary oxygen mask will provide an inspired concentration of around 50–60 % only, because the rest of the air inspired is entrained from the atmosphere, diluting the administered oxygen. When using nasal prongs, we normally administer oxygen at a rate of 3–5 L/min, which supplements inspired oxygen concentration to about 25–30 % depending on the state of opening of the mouth (breathing with open mouth allows more atmospheric air the be entrained into the airway, diluting the administered oxygen more).

Therapeutic Uses

Oxygen is used to treat hypoxia due to several causes: low inspired oxygen concentration (e.g. high altitude sickness); shunting in the lungs (e.g. pneumonia) and competition for binding with the hemoglobin molecule (e.g. carbon monoxide poisoning).

Hyperbaric oxygen (available from hyperbaric chambers) which increases the dissolved oxygen in the blood is useful for improving delivery of oxygen to the tissues, in conditions such as non-healing foot ulcers and osteoradionecrosis in cancer therapy. In decompression sickness, oxygen is administered under hyperbaric conditions and is used to replace bubbles of nitrogen gas in the blood which form following rapid ascent after diving. It is also used to treat air embolism.

Issues of Concern

Oxygen supports combustion. When administered in the hospital setting, patients should be advised to avoid smoking.

Patients with chronic obstructive airway disease rely on the hypoxic drive to initiate respiration. Administration of oxygen may depress respiration in these patients.

Normal oxygen saturations with oxygen therapy may delude a provider into thinking that all is well with the patient without managing the underlying cause of the problem.

Oxygen toxicity occurs when too much oxygen is administered either at a higher atmospheric pressure or at higher concentrations. At higher atmospheric pressure of oxygen, convulsion is a risk. With prolonged oxygen therapy, lung fibrosis can be a complication. Hypoxia at the tissue level on the other end causes harm too. It is recommended that we should aim for precise control of arterial oxygenation (Fig. 28.2) just as we do with blood glucose levels.

In preterm infants, there has always been concern about oxygen therapy resulting in retinopathy. Hyperoxia inhibits vascular endothelial growth factor in the first phase of retinopathy resulting in obliteration of retinal vessels and in the second phase there is pathological neovascularisation, which can lead to blindness.

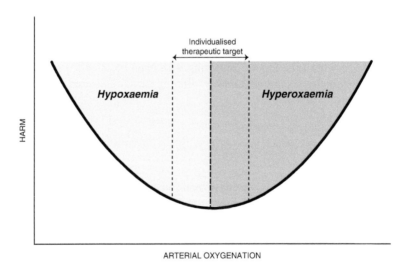

Fig. 28.2 Both hyperoxemia and hypoxemia cause harm so providers must aim for arterial oxygenation that precisely meets the patient's needs (With permission Wolters Kluwer; Crit Care Med 2013;41:423–32)

Clinical Correlation – What Level of Tissue Oxygen to Strive for in the Preterm Infant

Most providers are cautious about administering oxygen to preterm babies due to the possibility that too much oxygen will cause retinopathy of prematurity.

The study in NEJM 2013;368:2094–104, shows that premature babies ventilated to tissue oxygen saturation less than 90 % were 1.45 times more likely to die compared to their counterparts supported to saturation of more than 90 %. They were also 1.31 times more likely to have necrotising enterocolitis compared to their better oxygenated counterparts. The lower oxygenated preterm babies were 0.79 times less likely to get retinopathy of prematurity however.

Monitoring of Patients on Oxygen Therapy

Most acutely ill patients are monitored with the pulse oximeter, which shows the oxygen saturation of capillary blood through a probe placed on the finger. Normal saturation is 97–100 %. Although the oxygen dissociation curve is sigmoid shaped (Fig. 28.3), there is a linear relationship between oxygen saturation and oxygen tension at the critical range of poor oxygen reserve. This serves as an extremely useful non invasive alert of poor oxygenation and impending life threat which can

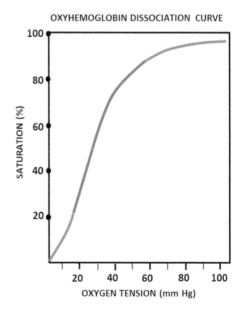

Fig. 28.3 The sigmoid nature of the oxygen dissociation curve reflects the non linear affinity of oxygen for the haemoglobin molecule. Between 90 and 20 % saturation (*red* portion of the curve) there is a roughly linear relationship between oxygen saturation and oxygen tension which allows the pulse oximeter to flag patients at risk of tissue hypoxia

Fig. 28.4 This shows the relationship between the oxygen tension in a patient and the inspired oxygen concentration reflecting the degree of shunting in the lungs (With permission from Springer; Respiratory effects of general anaesthesia – ebook & Oxford University Press; Br J Anaesth 1973;45:711)

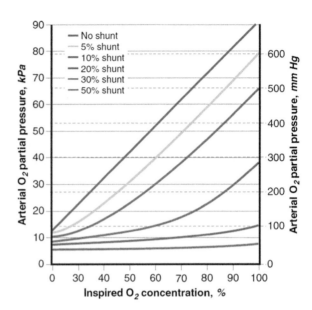

be picked up by the pulse oximeter. Increasingly, the cerebral oxygen monitor is also being used to monitor the global cerebral oxygen saturation.

The arterial blood gas is a better monitor of the state of oxygenation at the tissue level. The various parameters include the pH (7.35–7.45); pO_2 (95–100 mmHg); pCO_2 (35–45 mmHg); Bicarbonate (20–24 mmol/L); Base excess (+/−2) and Lactate (<1.5 mmol/L). A rising lactate and base excess suggest that anaerobic respiration is occurring at the tissue level.

The oxygen tension when determined with the FiO_2 gives an indication of the degree of shunting occurring in the lungs (Fig. 28.4). Alternatively we can use the P/F ratio where the oxygen tension (represented by P mmHg) is divided by the fractional oxygen concentration (represented by F) delivered to the patient. A ratio of less than 300 indicates significant shunting in the lungs.

Key Concepts

- Availability of oxygen at the tissue level allows the more efficient production of ATP from energy sources via aerobic metabolism.
- Oxygen moves into the tissue from the environment through the cardio-respiratory system and is mainly transported by hemoglobin. Intact functioning of these systems is essential to ensure adequate oxygenation at the tissues.
- Oxygen is administered to overcome inadequate delivery to the tissue caused by dysfunction of the cardio-respiratory system or deficit in oxygen carriage by hemoglobin.

- Oxygen is important to treat hypoxia but it carries hazards in the form of fire risks, depression of respiratory drive, respiratory fibrosis with prolonged use and seizures in those administered high barometric pressures of oxygen.

Summary

Oxygen is a drug naturally available in the atmosphere, but given in higher concentrations from special oxygen sources to the acutely ill patient. It is administered via nasal prongs, masks or other oxygen delivery and/or ventilator devices. Oxygen is delivered to the tissues by the cardiopulmonary system and is mainly carried by hemoglobin. Oxygen improves the efficiency with which ATP can be produced from energy sources at the tissue level. It is utilised to treat hypoxia from any cause. Excessive oxygen at the tissue level is associated with tissue damage e.g. retinopathy of prematurity, lung fibrosis and seizures. Oxygen administration areas are a fire hazard risk.

Further Reading

1. Benatar SR, Hewlett AM, Nunn JF. The use of iso-shunt lines for control of oxygen therapy. Br J Anaesth. 1973;45:711–8.
2. Brodie D, Bacchetta M. Extracorporeal membrane oxygenation for ARDS in adults. N Engl J Med. 2011;365:1905–14.
3. Costa EL, Amato MB. The new definition for acute lung injury and acute respiratory distress syndrome: is there room for improvement? Curr Opin Crit Care. 2013;19(1):16–23.
4. Davis PG, Tan A, O'Donnell CPF, Schulze A. Resuscitation of newborn infants with 100 % oxygen or air: a systematic review and meta-analysis. Lancet. 2004;364:1329–33.
5. Elliot S. Erythropoiesis – stimulating agents and other methods to enhance oxygen transport. Br J Pharmacol. 2008;154:529–41.
6. Leach RM, Treacher DF. The pulmonary physician in critical care. 2: oxygen delivery and oxygen consumption in the critically ill. Thorax. 2002;57:17–77.
7. Martin DS, Grocott MP. Oxygen therapy in critical illness: precise control of arterial oxygenation and permissive hypoxemia. Crit Care Med. 2013;41:423–32.
8. Martin DS, Grocott MP. Oxygen therapy in anaesthesia: the yin and yang of O_2. Br J Anaesth. 2013;111(6):867–71.
9. The BOOST II United Kingdom, Australia, and New Zealand Collaborative Groups. Oxygen saturation and outcomes in preterm infants. N Engl J Med. 2013;368:2094–104.

Chapter 29
Fluids as Drugs

Marzida Mansor

Abstract Oxygen and energy sources are brought to the tissues in the flow of blood. Fluids are the vehicle through which these essential life-sustaining molecules are transported and moved. The use of fluid resuscitation with colloid and crystalloid solutions is a universal intervention in the treatment of the acutely ill patient. However, it is now well accepted that no ideal resuscitation fluid exists. The selection and use of resuscitation fluids is based on physiological concepts, but clinical practice depends largely on clinician preference, training and geographical differences. Nevertheless, there is strong evidence from major clinical trials that the choice of the type and dose of resuscitation fluid may affect patient outcome. Hence, clinicians should prescribe the resuscitation fluids as they would prescribe any other intravenous drug. The selection of the specific fluid should be based on indications, contraindications, accurate dosing and potential toxicities in the hope of maximizing efficacy and minimizing iatrogenic complications. The optimal dosing should involve assessment of volume responsiveness and the context in which the fluid is being used.

Keywords Fluid resuscitation • Colloid • Crystalloid • Balanced solutions • Mortality • Renal replacement therapy • Toxicity

Introduction

The use of intravenous fluids for resuscitation and maintenance has been widely discussed over many years. The wide variety of crystalloids and colloids available in the market has further complicated decision making. The ideal resuscitation fluid should be one that produces a predictable and sustained increase in intravascular volume, has a chemical composition as close as possible to that of extracellular fluid, does not produce adverse metabolic or systemic effects, is metabolized and completely excreted without accumulation in tissues, produces good patient outcomes and is cost-effective. To date, there is no ideal fluid available for clinical use.

M. Mansor, M.D., M.Anaes (✉)
Department of Anesthesiology, Faculty of Medicine, University of Malaya,
50603 Kuala Lumpur, Malaysia
e-mail: marzida@ummc.edu.my

© Springer International Publishing Switzerland 2015   239
Y.K. Chan et al. (eds.), *Pharmacological Basis of Acute Care*,
DOI 10.1007/978-3-319-10386-0_29

However, there is new evidence that the type and dose of fluids used in resuscitation will affect patient outcome. The use of colloids such as albumin is not only costly but is associated with an increase in mortality among patients with traumatic brain injury. The use of hydroxyethyl starch (HES) solutions among intensive care patients is associated with increased rates of renal-replacement therapy and adverse events. On the other hand, the use of crystalloids such as normal saline has been associated with development of metabolic acidosis and acute kidney injury.

The complications of under-resuscitation with unbalanced fluids include tissue hypoxemia, risk of acute kidney injury, lactic acid and unmeasured anion acidosis, and gastrointestinal disturbances, while over-resuscitation may result in tissue edema causing hypoxia, compartment syndrome and renal dysfunction, hyperchloremic metabolic acidosis and risk of hypernatremia, anastomotic leakage, diarrhea and other gastrointestinal disturbances, pulmonary edema, hepatic congestion and prolonged mechanical ventilation. There is increasing evidence that balanced salt solutions are superior to unbalanced ones, therefore, it is pragmatic to use them as the initial resuscitation fluid.

In light of the above findings, it is recommended that fluids be administered with the same caution that is used with any other intravenous drug. When administering fluids, consider the type, dose, indications, potential for toxicity and the cost.

Physiology of Body Fluids and Electrolytes

The selection of the specific fluid for resuscitation has always been based on physiological concepts which are explained by the fluid compartment model and the factors that dictate fluid distribution across these compartments. The total body water (TBW) in a 70-kg man is about 42 L. This constitutes 60 % of the total body weight.

The water is distributed into two main compartments in the body (Fig. 29.1):

- Intracellular fluid (ICF) compartment – 2/3 of the 42 L TBW (28 L)
- Extracellular fluid (ECF) compartment – 1/3 of the 42 L TBW (14 L)

The ECF can be further subdivided into:

- Interstitial fluid compartment, water in dense connective tissue and water of bone and transcellular fluid compartment (ISF) – 75 % of ECF
- Intravascular fluid compartment (Plasma) – 25 % of ECF

The transcellular fluid compartment consists of a collection of fluids, formed during the process of the transport activities in cells and is found mainly in epithelial lined spaces. They are important because of the special roles they play in the body. Examples of transcellular fluids are cerebrospinal fluid, joint fluid, fluid in the bowel, fluid in body cavities, aqueous humor, bile in the gall bladder and urine in the bladder.

Fig. 29.1 Body fluid
compartments in terms of
volumes of fluid

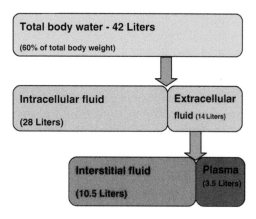

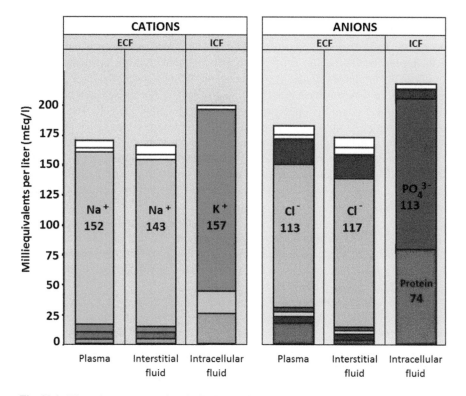

Fig. 29.2 Electrolyte concentrations in ECF and ICF

All of these fluids contain electrolytes (ions) that are distributed in such a way as to maintain electrical neutrality (Fig. 29.2). Potassium is the most abundant electrolyte in the ICF and sodium in the ECF. The difference is maintained by the basolateral Na^+/K^+ ATPases, which transport three Na^+ ions out of the cell in exchange for two K^+ ions transported into the cell. A balance of charges is maintained in each compartment by different ions.

Major fluid and electrolyte shifts occur daily between the various fluid compartments. Some of these fluid shifts require energy, others do not. In the acutely ill patients, these shifts are even more exaggerated.

Water moves passively across semi-permeable membranes from areas of lower to higher osmolality by a process called osmosis. Osmolality refers to the concentration of osmotically active particles, like proteins, expressed in terms of osmoles of solutes per kilogram of solvent. The ICF is hypertonic, or hyperosmolar, compared to the interstitial fluid, so water will move into the cell by osmosis. This will occur until the osmolality is the same on both sides of the membrane. This may explain the higher proportion of intracellular water compared to extra-cellular water (28 vs 14 L in a 70-kg man). If the osmolality of the ECF were increased (as in dehydration), then there would be net water movement out of cells and this would continue until the osmolality in the ICF equals that of the ECF. However, as the osmotic pressure between the intracellular compartment and the interstitial space is minimal, there is very little fluid movement between these two areas.

In normal healthy cells, active transport maintains a high concentration of potassium and phosphate and a low concentration of sodium within the cell. There is thus a constant tendency for these electrolytes (i.e., potassium and phosphate) to diffuse down the concentration gradient out from the cell. The sodium-potassium pump, in an energy-requiring process, constantly pumps the potassium back into the cell in exchange for sodium, maintaining the electrolyte concentrations, resulting in a potential difference across the membranes of the cells, which is vital for cell function and life. As long as there is enough energy to maintain the function of the sodium-potassium pumps, the ionic distribution is maintained and there is homeostasis.

At the level of the capillary, oxygen, ions and nutrients (in the form of electrolytes, glucose, amino acids and lipids) pass from the blood to the interstitial fluid compartment where they subsequently move into the cells. Waste products (carbon dioxide and acids) from the cells move in the opposite direction from the cells back to the capillaries through the interstitial space.

Fluid movement at this site is determined by the differences in hydrostatic and oncotic pressures between the capillaries and interstitial fluid. These forces, also known as Starling forces, result in net movement of fluid into or out of the capillaries (Fig. 29.3). The normal fluid transfer occurs out of the capillaries and amounts to about 1 ml per 100 g tissue per minute.

The principal determinants of transvascular exchange, has always been thought to be due to the Starling forces across the capillaries. Recent findings have questioned these classic models. Non-fenestrated (continuous) capillaries (see Chap. 4, Fig. 4.2) throughout the interstitial space have been identified, indicating that absorption of fluid does not occur through venous capillaries but that fluid from the interstitial space, which enters through a small number of large pores, is returned to the circulation primarily as lymph that is regulated through sympathetically mediated responses.

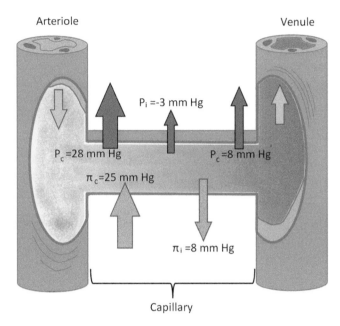

Fig. 29.3 Starling forces across the capillary. The Starling forces (hydrostatic [P_c and P_i in the capillary and interstitium respectively] and oncotic pressures [π_c and π_i in capillary and interstitium respectively]) allow for bulk flow of fluid and nutrients across the capillary wall

The endothelial glycocalyx layer, a web of membrane-bound glycoproteins and proteoglycans on the luminal side of endothelial cells has been identified as the key determinant of membrane permeability (Fig. 29.4). The subglycocalyx space produces a colloid oncotic pressure that is an important determinant of transcapillary flow. The structure and function of the endothelial glycocalyx layer can determine the membrane permeability of various vascular organ systems. The integrity, or "leakiness," of this layer, and thereby the potential for the development of interstitial edema, varies substantially among organ systems, especially under inflammatory conditions, such as sepsis, and after surgery or trauma. These are the conditions in which resuscitation fluids are commonly used.

Types of Resuscitation Fluid

Resuscitation fluids are broadly categorized into crystalloid and colloid solutions (Table 29.1). Colloid solutions are suspensions of molecules within a carrier solution that are relatively incapable of crossing the healthy semipermeable capillary membrane owing to the molecular weight of the molecules. Crystalloids are solutions of ions that are freely permeable but contain concentrations of sodium and chloride that determine the tonicity of the fluid.

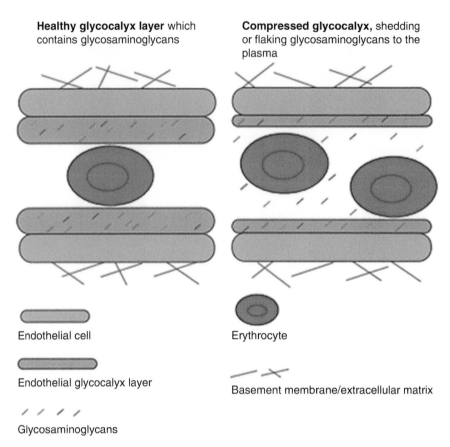

Fig. 29.4 The endothelial glycocalyx layer in healthy patients. Note the damaged endothelial glycocalyx layer in those whose microcirculation is impaired (Permissions from Oxford University Press; BJA 2012;108(3):369–71)

A Cochrane systematic review on colloids versus crystalloids for fluid resuscitation in critically ill patients revealed that there is no evidence from randomized controlled studies that resuscitation with colloids reduces the risk of death, compared to resuscitation with crystalloids in patients with trauma, burns or following surgery.

Despite this, proponents of colloid solutions continue to argue that colloids are more effective in expanding intravascular volume because they are retained within the intravascular space and maintain colloid oncotic pressure. The volume-sparing effect of colloids, as compared with crystalloids, is considered to be an advantage. Conventionally, the volume of crystalloids required to maintain intravascular volume is three times that of colloids. In addition, the use of crystalloids has classically been associated with the development of clinically significant interstitial edema. Semisynthetic colloids have a shorter duration of effect than human albumin solutions but are actively metabolized and excreted.

Table 29.1 Types and composition of resuscitation fluids available currently

Types and compositions of resuscitation fluids

Variable	Human plasma	Colloids								Crystalloids		
		Albumin 4 %	Hydroxyethyl starch					Succinylated modified fluid gelatin 4 %	Urea-linked gelatin 3.5 %	0.9 % Saline	Compounded sodium lactate	Balanced salt solution
			10 % (200/0.5)	6 % (130/0.4)	6 % (130/0.4)	6 % (130/0.42)	6 % (130/0.42)					
Trade name		Albumex	Hemohes	Voluven	Volulyte	Venofundin	Tetraspan	Gelofusine	Haemaccel	Normal saline	Hartmann's or ringer's lactate	Plasmalyte
Colloid source		Human donor	Potato starch	Maize starch	Maize starch	Potato starch	Potato starch	Bovine gelatin	Bovine gelatin			
Osmolarity (mOsm/L)	291	250	308	308	286	308	296	274	301	308	280.6	294
Sodium (mmol/L)	135–145	148	154	154	137	154	140	154	145	154	131	140
Potassium (mmol/L)	4.5–5.0				4.0		4.0		5.1		5.4	5.0
Calcium (mmol/L)	2.2–2.6											
Magnesium (mmol/L)	0.8–1.0											
Chloride (mmol/L)	94–111	128	154	154	110	154	118	120	145	154	111	98
Acetate (mmol/L)					34		24					27
Lactate (mmol/L)	1–2										29	
Malate (mmol/L)							5					
Gluconate (mmol/L)												23
Bicarbonate (mmol/L)	23–27											
Octanoate (mmol/L)		6.4										

Proponents of crystalloid solutions have argued that colloids, in particular human albumin, are expensive and impractical to use as resuscitation fluids. Furthermore the use of HES in critically ill patients might increase mortality. Crystalloids are inexpensive and widely available and have an established role as first-line resuscitation fluids.

Determination of Dose and Volumes Needed

The key concept for fluid dosing using crystalloids in critically ill patients is to actively address ongoing losses and to assess the need for circulatory support. 'Routine maintenance' or replacement of unmeasured presumed fluid deficits needs to be repeatedly questioned. Threshold central venous pressure (CVP) values continue to be advocated as targets for crystalloid bolus resuscitation in mechanically ventilated and spontaneously breathing populations even in recently updated guidelines. Such CVP-guided therapy may be harmful and should be abandoned in favor of objectively defined individualized goals integrating functional hemodynamic parameters. Several parameters that are superior to blood pressure, CVP, and urine output targets are available, including pulse pressure variation and variations in sonographic dimensions readily assessed with limited training at the bedside.

When conventional urine output or blood pressure targets guide volume therapy in severely ill patients, large doses of fluids result and HES appears especially harmful. The use of boluses in critically ill patients without systematic examination of objective functional parameters appears to be harmful. Holding dose constant, differences between fluid types depend on the specific toxicity being measured and on underlying patient severity of illness. Important patient-dependent and context specific differences need to be considered.

Key Concepts

- The ideal resuscitation fluid should be able to increase intravascular volume without adverse effects and it should be cost effective.
- There are major fluid shifts across compartments due to osmolality differences and this is a consideration when we are administering fluids to meet patients' needs.
- The endothelial glycocalyx plays an important role in the integrity of the intravascular compartment. In some situations, e.g., sepsis and trauma, the integrity of the glycocalyx is impaired, so fluid resuscitation becomes complicated.
- Fluid therapy in excess of what is needed by patients can be harmful.
- Adequacy of fluid therapy guided by new monitoring techniques like pulse pressure variations may be superior to conventional methods.

Summary

Fluid balance is a component of a complex physiological process. It is important to identify the fluid that is most likely to be lost and replace the fluid lost in equivalent volumes. Fluid requirements may change over time in the acutely ill patient and hence it is necessary to keep up with these requirements. In selecting the type and dose of intravenous fluids, specific considerations should be applied to different categories of patients.

Further Reading

1. Myburgh JA, Mythen MG. Resuscitation fluids. N Engl J Med. 2013;369:1243–51.
2. Wujtewicz M. Fluid use in intensive care. Anaesthesiol Intensive Ther. 2012;44(2):92–5.
3. Krogh A, Landis EM, Turner AH. The movement of fluid through the human capillary wall in relation to venous pressure and to the colloid osmotic pressure of the blood. J Clin Invest. 1932;11:63–95.
4. Levick JR, Michel CC. Microvascular fluid exchange and the revised Starling principle. Cardiovasc Res. 2010;87:198–210.
5. Woodcock TE, Woodcock TM. Revised Starling equation and the glycocalyx model of transvascular fluid exchange: an improved paradigm for prescribing intravenous fluid therapy. Br J Anaesth. 2012;108(3):384–94.

Part V
Safety Issues in Pharmacology

Chapter 30
Medication Errors

Kwee Peng Ng

Abstract Medication errors are a subset of all medical errors that patients are exposed to upon a visit to hospital. Errors related to drugs do not all result in an adverse drug event and not all adverse drug events are the result of an error. However, all medication errors are preventable. They can occur during any phase of the medication of a patient, from the prescription to the actual ingestion or administration of the drug to the patient. The incidence of medication and medical errors is difficult to determine, due to differences in practices in hospitals worldwide, and frequency of reporting. Nonetheless, errors do occur and so it is incumbent upon the medical community, in the interest of patient safety, to be aware of the problem and be constantly vigilant to keep errors at a minimum. A systems based approach is part of the strategy to address medication errors in hospitals.

Keywords Medication error • Adverse drug event • Patient safety • Systems based approach

Introduction

Medications are administered to patients to cure and give relief, but unfortunately can also be the cause of unintended pain and dis-ease in patients. Adverse drug events have the potential to occur whenever drugs are prescribed and a proportion of these adverse events are preventable. Preventable adverse drug events are caused by mistakes, and are collectively referred to as medication error. Medication errors can occur at any stage in the process of medicating a patient, namely during the prescription, transcription, dispensing and administration of drugs. The result of these errors can range from absolutely no harm caused, to major life threat or even death.

The occurrence of errors in healthcare was brought to the attention of all stakeholders by the Committee on Quality of Health Care in America in 2000 in a landmark publication - "To err is human: Building a Safer Health System". In this publication, it was reported that adverse events occurred in 2.9 and 3.7 % of hospitalizations in 2 separate surveys done, and that medication error was

K.P. Ng, M.B.B.S., M.Anaes, FANZCA (✉)
Subang Jaya Medical Centre, 47500 Subang Jaya, Malaysia
e-mail: drkpng@hotmail.com

© Springer International Publishing Switzerland 2015
Y.K. Chan et al. (eds.), *Pharmacological Basis of Acute Care*,
DOI 10.1007/978-3-319-10386-0_30

responsible for 19 % of adverse events occurring in hospitalized patients. The extent of the problem was further quantified by MEDMARK, an internet accessible medication error reporting program in the United States, which revealed that medication error occurred in 78 out of every 100,000 emergency department visits in a 2008 publication. Accurate data on the frequency of drug errors are difficult to determine, due poor reporting mechanisms and the prerequisite voluntary nature of reports. Frequencies of medical and medication errors vary widely due to different hospital practices and work cultures. Lack of data in many institutions makes it difficult for the problem to be properly addressed, for patient safety measures to be developed and for the outcome of these measures to be analyzed.

Medication Delivery Process

There are many processes involved before the medication ordered is actually administered to the patient. During this delivery process, medication errors can occur and the frequency with which this occurs is highlighted in Table 30.1.

1. Prescribing
   This involves the assessment for the need to prescribe a drug and the selection of the said drug.
2. Transcribing
   This occurs in the traditional health care setting where the prescription written by the doctor is read and transcribed in the pharmacy.
3. Dispensing
   At this stage of medication of patients, the prescription is processed in the pharmacy and the drug is prepared and dispensed to the patient/wards in the correct dosage, form and at the correct time.
4. Administration
   Once the drug is received in the wards, the right drug must be administered to the right patient in the right way. If possible, patients should be informed about the medication they receive and educated about the administration of the drug.
5. Monitoring

Table 30.1 Process during which medication errors are likely to occur

Node where error occurred	Actual errors (n = 11,997)
Prescribing	29 %
Transcribing/documenting	25 %
Dispensing	8 %
Administering	36 %
Monitoring	1 %

Permission from Elsevier; the Journal of Emergency Medicine 2011;40(5):485–92

Patient's response to drugs should be monitored and documented. Any adverse drug reaction should be identified and reported and the drug therapy should be reevaluated and altered if necessary.

Factors That Contribute to Errors

The more frequent and numerous the prescription of drugs in a particular setting, the more the risk of an error occurring. It is well documented that most medication errors occur in Intensive Care Units, Pediatric Intensive Care Units (because of the wide ranges of patients in terms of age and weight, and requirement for weight determined dosing), and in the elderly patients.

In the prescribing of drugs, knowledge about the drug is a prerequisite to enable proper drug and dosage order. Adequate information about the patient is necessary (e.g. history of drug allergy) to avoid the use of the drug concerned. Drug dose must be adjusted for patients with disease states like hepatic and renal dysfunction.

The written prescription is usual in traditional practice, so illegibility and error prone methods of writing dosages are rife (e.g. writing of 1.0 mg can easily be mistaken as 10 mg, and use of wrong units). Abbreviations are also factors in drug errors – q.d. or once daily can easily be mistaken for q.i.d. which is four times daily. Providers must be particularly mindful of errors due to drugs with similar sounding names (e.g. Zyrtec and Zantac) and drugs which look alike or are put into look-alike packaging.

After transcription of the written prescription, the drugs are dispensed. This involves accurate compounding of the drug in the proper dose, form and timing. Drugs can be incorrectly formulated at this stage.

Administration errors occur when drugs are dispensed to the wrong patient, in an inappropriate dose or form and at the wrong time. These errors are generally human errors which tend to occur more frequently in busy units where nurses are stressed, overworked or inexperienced, as well as in the operation theatres when anesthetists often administer multiple drugs in often highly stressful situations.

The response to a drug must be monitored by providers with the assistance of the patient so that in the event that an adverse reaction does occur, appropriate adjustment or change in the drug administration can be effected. Documentation of such adverse drug effects is important to avoid similar errors in future.

Preventing Medication Errors

The publication of the eye-opening paper "To err is Human" spurred efforts to minimize the occurrence of medication errors (as well as medical errors) through establishment of organizations to look into and manage every aspect of patient safety, such as Centers for Patient Safety and Institute for Safe Medication

Practices. Error reporting processes that were voluntary and confidential were set up to determine the full extent of the occurrence of errors, thus enabling steps to be taken to address factors responsible for them. Information was shared and experts helped identify best practices to be implemented. Regulations and guidelines were formulated to improve safety compliance and training programmes conducted to train personnel in safety issues.

A systems based approach was established to improve conditions of patient medication, to make it difficult for errors to occur. Basically the environment in which health care is provided and the people working within it interact as a complex and dynamic functioning unit or system. Parts of the system relate to each other directly or indirectly. All systems have processes that transform inputs (e.g. ill patients) into outputs (e.g. healthy individuals). So, systems based approaches allow us to focus on the processes in the system and to use key target measures to evaluate and improve these processes in the system.

Computerized entry of prescriptions abolished the errors occurring from illegibility and transcription. Potential adverse drug reactions may be prevented by pharmaceutical software which enables automated checking for drug doses, allergies and drug interactions. Bar coding can additionally be implemented to minimize wrong patient dosing. High alert medications should be identified and flagged as such, as these drugs have the potential to cause significant harm if administered wrongly. High risk drugs should not be stored in patient care areas with easy access to them.

The clinical pharmacist has a huge role in reducing errors especially during the dispensing of medications in high risk areas. Clinical pharmacists working in intensive care units can not only ensure the correct dose and timing of drugs prescribed, they can greatly reduce errors of drug-drug interactions or allergy risk of inadvertent prescription of drugs of the same class.

Administration of drugs following strict protocols by nurses has been established. The 'Five Rights' of patients, i.e. the right to the administration of the right drug, in the right dose at the right time by the right route to the right patient has been enforced in some centers. High alert medication administration must be checked by two nurses prior to administration. Look alike and sound alike medications should raise red flags and if possible, drugs in look-alike packaging should be eliminated from use. Nurses tasked with administration of medications should not be distracted during the activity. Infusion pumps used to administer drugs should have relevant safety features and care givers should be proficient in their use. Table 30.2 gives an idea of the causes and factors contributing to medication errors in a busy emergency department.

Apart from systems based practices, professional issues determined by the individual practitioners' skill, knowledge, competence and experience obviously play a role in the provision of safe health care. Performance standards of health care givers should be determined and regularly evaluated and assessed. Errors committed due to lack of knowledge are avenues for improvement strategies. There is a growing realization that there is a disconnection between the teaching of pharmacology, imparting knowledge about pharmacokinetics and pharmacodynamics of

Table 30.2 Causes and contributing factors of medication errors

Cause and contributing factor	Reported errors (n = 13,932)
Cause	
Procedure/protocol not followed	17 %
Communication	11 %
Abbreviations	5.6 %
Transcription inaccurate/omitted	4.1 %
Calculations errors	3.8 %
Dispensing device involved	3.4 %
Contraindicated, drug allergy	3.0 %
Computer entry	2.5 %
Handwriting illegible/unclear	2.3 %
Written order	5.7 %
Verbal order	3.7 %
Performance deficit	29 %
Knowledge deficit	9.2 %
Contributing factor	
Distractions	7.5 %
Emergency situation	4.1 %
Workload increase	3.4 %
Staff, inexperienced	3.1 %
Patient transfer	2.2 %
Cross coverage	1.5 %
No 24 h pharmacy	1.4 %
No access to patient information	1.4 %
Shift change	1.0 %

Permission from Elsevier; the Journal of Emergency Medicine 2011;40(5):485–92

drugs during the preclinical education, and the actual ability of young doctors to safely prescribe drugs. Suggestions have been made for the teaching of clinical pharmacology to be incorporated in the teaching of medical students, particularly at the bedside to maximize the impact of this teaching. Many medical schools revisit the subject of pharmacology during the clinical years in order to address this.

Patients have a role to play in the prevention of adverse drug events as well. Patient education and comprehension regarding the drugs they are taking especially in the outpatient setting can help intercept errors such as wrong drug or dosage. Patients should be cognizant of their allergy history and possible drug interactions particularly if they are taking multiple drugs. Supervised drug administration is probably advisable in the old and infirm who are unable to cope with multiple drug dosing for frequently multiple medical conditions.

The Cost of Medication Errors

The costs of medication errors are not only financial but also social. Adverse drug effects result in prolongation of medical care, additional treatment, increase in number of investigations ordered and more visits to emergency rooms and clinics. Although most medication related adverse events do not result in harm to the patient, those that do are often costly. A large prospective study determined that about 12 % of adult emergency department visits were due to symptomatic adverse drug effects, of which 40–50 % was not recognized by the emergency room physicians. Loss of manpower hours, productivity and disability payments are associated indirect costs.

In the social sense, erosion in the confidence in the healthcare system and decreased patient satisfaction result. It is the responsibility of all healthcare providers to ensure the safety of those under their care. In this context, safety has been defined as freedom of the patient from accidental injury.

Medication Errors in Perspective

Statistics vary and are not available from many centers. Errors with potential to cause harm have been reported to occur in up to 2.6 % of all medications ordered. Underreporting of adverse drug effects probably exists. Ideally, measures undertaken to address the issue of medication errors should totally prevent them. In reality this is impossible. Medication errors can be reduced through system orientated approaches, improvements in the education of our healthcare providers and patients, and being true to the Hippocratic Oath of doing no harm. While many changes have been instituted to improve patient safety, unfortunately the actual result has been less than what has been hoped for. The reasons for this are many and complicated, including fear of litigation, the complexity of the healthcare industry and financial issues. However, an important positive outcome of the report "To err is human" is the change in the attitudes of healthcare professionals and organizations. They now realize that many medical errors (a large part being medication errors) are preventable and view the threat to patient safety seriously. Healthcare authorities and industry members support patient safety through regulations and guidelines. Accrediting agencies such as the Joint Commission International have made patient safety a priority in their global healthcare standards. It is anticipated that with the continuing efforts of all stakeholders, a safer healthcare system will come into being.

Key Concepts

- Medication errors are a source of patient morbidity and mortality.
- Medication errors can occur at any stage of the medication delivery system.

- A systems based approach that makes it difficult for errors to be committed has been established as part of the efforts to ensure patient safety.
- Administrators, healthcare providers and patients all have a role to play in reducing medication errors.

Summary

Medication errors make up a significant part of medical errors, which are defined as injuries suffered by patients as a result of medical management which are preventable. These errors as a whole contribute significantly to patient mortality and morbidity which not only cause pain and suffering, but also financial loss. Current efforts to improve patient safety in the healthcare system have yet to show major results but very importantly, the attitudes of healthcare professionals and organizations involved in healthcare have changed. There is now awareness of the extent of the occurrence of medical errors, the possibility of preventing them and the will and methods to do so. Systems based practices and approaches to decrease medical errors are being adopted worldwide as part of the solution in protecting the safety of our patients.

Further Reading

1. Etchells E, Juurlink D, Levinson W. Medication errors: the human factor. Can Med Assoc J. 2008;178(1):63–4.
2. Gokhman R, Seybert AL, Phrampus P, Darby J, Kane-Gill SL. Medication errors during medical emergencies in a large, tertiary care, academic medical centre. Resuscitation. 2012;83:482–7.
3. Institute of Medicine. To err is human: building a safer health system. Washington, DC: The National Academies Press; 2000.
4. Lehman DF. Teaching from catastrophe: using therapeutic misadventures from hydromorphone to teach key principles in clinical pharmacology. J Clin Pharmacol. 2011;51(11):1595–602.
5. Pham JC, Story JL, Hicks RW, Shore AD, Morlock LL, Cheung DS, et al. National study of the frequency, types, causes and consequences of voluntarily reported emergency department medication errors. J Emerg Med. 2011;40(5):485–92.

Index

© Springer International Publishing Switzerland 2015
Y.K. Chan et al. (eds.), *Pharmacological Basis of Acute Care*,
DOI 10.1007/978-3-319-10386-0